Professional Development through Action Research in Educational Settings

Professional Development through Action Research in Educational Settings

Edited by

Christine O'Hanlon

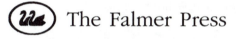 The Falmer Press

(A member of the Taylor & Francis Group)
London • Washington, D.C.

UK   Falmer Press, 1 Gunpowder Square, London, EC4A 3DE
USA   Falmer Press, Taylor & Francis Inc., 1900 Frost Road, Suite 101, Bristol,
        PA 19007

First published in 1996

**A catalogue record for this book is available from the British
Library**

**Library of Congress Cataloging-in-Publication Data are
available on request**

ISBN 0 7507 0507 8 cased
ISBN 0 7507 0508 6 paper

Jacket design by Caroline Archer

Typeset in 10/12 pt Garamond by
Graphicraft Typesetters Ltd., Hong Kong.

*Printed in Great Britain by Biddles Ltd, Guildford and King's Lynn
on paper which has a specified pH value on final paper manufac-
ture of not less than 7.5 and is therefore 'acid free'.*

Contents

Contents

List of Figures and Tables

Acknowledgments

Both the Editor and the Publisher would like to acknowledge that permission has been given to reproduce the following articles:

Peter Posch 'Changes in the Culture of Teaching and Learning and Implications for Action Research', *Education Action Research Journal*.

Melanie Walker 'Developing the Theory and Practice of Action Research: A South African Case', *Education Action Research Journal*.

John Elliott and Clem Adelman (1973) Reflecting Where the Action Is: The Design of the Ford Teaching Project', in Education for Teaching, *Journal of the Association of Teachers in Colleges and Departments of Education*, **92**, pp. 8–20, Autumn.

Introduction

This book offers readers an international perspective on the role of action research or practical enquiry in educational contexts. It is a selection of papers from a CARN (Collaborative Action Research Network) Conference held at the University of Birmingham in the Spring of 1994. One exception is the paper written by John Elliott and Clem Adelman which has been included because of its significance, now retrospectively, in 'reflective practice' in a UK context in the 1970s, in view of the work of Schön in the USA in the 1980s. The chapters written by Hollingsworth, Walker and O'Hanlon were delivered as keynote addresses during the conference. The book is divided into three sections to reflect the variety of interpretations and working contexts in which action research plays a role.

In Part I, John Elliott and Clem Adelman's paper, although written originally more than twenty years ago, illustrates how practical enquiry has supported autonomous student learning, evidence-based deliberation, and reflection on action. Furthermore, it brings about the modification of practice and collaborative decision making in schools. The themes in this paper indicate to readers the significance of the Ford Teaching Project and the Humanities Curriculum Project in initiating the development of the 'reflective practitioner'. In the UK in the 1970s before Schön's book with that title was published in the USA in 1983. (Schön 1983). It also provides a basic example for understanding how action research operates in practice. The themes expanded in this paper are echoed throughout the book in the contributions from the UK, USA, Austria, Australia and South Africa.

Sandra Hollingsworth's contribution to the conference was immense because of her special delivery and the message she brought to it. The political empowerment of her students, their increasing awareness of themselves as activists attempting to change the world and the pain and potential of such analyses are underlying themes for the action research. She describes how feminist postmodern contributions to eroding the 'theorizing subject' have left her with a sense of agency. She argues towards the proposition that feminist notions of self-positioning, self-critique and continuous challenge to (even our own) constructed metanarratives, offer a strong reinterpretation of action research in both university and school organizations and practices. The naming of shifting subjectivities and political powers controlling us can help us to overcome their disempowering effects.

Melanie Walker's article puts action research into another political context, that of the newly politically reconstituted South Africa. Her article reflects themes from Hollingsworth in its autobiographical insights into how life experience influences one's understanding of the local political research and educational policy climate. Her own educational development is grounded in the dialectical play of sociological theories and empirical texts for change through political action and action research. She uses her own experience and the experiences of her colleagues involved in action research projects to critique the notion of action research with a plea for rigorous theory based projects to advance educational learning from personal experience and practical research. She concludes by putting change in the political context of South Africa, and indicates what that means to her and for collective action in action research projects.

Peter Posch puts action research into a global culture of change in teaching and learning. He examines the challenges for education and the necessity of using action research to cope with 'dynamic developments, divergent demands to complex practical situations'. These examples of a learning culture in the future feature a pan-European view of how schools are being restructured and a view drawn from Austria, Scotland and Sweden around the theme of environmental education. He refers back to the Humanities Curriculum Project framework as an example of 'reflective conversations' as per Schön (1987). His thesis is based around the proposition that as the heterogeneity of demands on the person/teacher increases in industrialized society, so too will the executive, organizational and supervisory practice with organizations increase. Part I reflects a range of national contexts within which action research is employed for different ends. Its practice, definition, profile and effects on teachers and students are re-defined and presented for further discourse and analysis.

Part II examines how action research is being used to develop specific practice in particular identified contexts, i.e., in Quality Development projects in schools, in a campaign for inclusive education, to improve cooperative learning and thinking skills with students, and to stimulate change in school for professionals working with emotionally and behaviourally disturbed children.

Christine O'Hanlon's paper compares quality development initiatives in education with action research which has an educational purpose. The quality development process uses a specific technical management approach to change through research and evaluation which omits a number of crucial factors integral to the action research process which make it intellectually challenging, empowering and ultimately educational. The quality development process is outlined, its strengths and weaknesses are identified and a new critique is made of how a critical educational process could achieve systematic re-framing of school-based research and evaluation for alternative forms of practice. The paper puts particular stress on the importance of self-knowledge as a means of knowing in evaluation which is particularly useful when teachers may be working in a school environment of compliance and coercion.

Alice Paige-Smith presents an action research case study of a parent and professional campaign for the retention of inclusive education in a London borough. She discusses critically the dilemmas of conducting action research with parents in her role as a researcher, parent and advocate, and how this leads to a conflict of identity. A diary of events was used to collate the information in the case study which raised questions about its use. She became an observer, an active participant and a recorder of events which enabled her to treat the parents as active subjects empowered to understand and change their situations.

Jane Richards uses an action research process to examine the teachers role in redefining and reconstructing cooperative learning in groups, and their learning to think more deeply and challenge other writers' opinions and thinking. She describes how she develops this process over two years and how she evaluates its success. Her reflections shows insight into the benefits and drawbacks in her activities and she concludes with implications for student teaching and its relationship to team building in the workplace.

Sonia Burnard and Heather Yaxley illustrate the dynamic nature of action research when teachers are committed to a constructive process of change. Their chapter focuses on the ability of action research to stimulate change, ideas and construction within a school programme. It not only reflects the thinking and active involvement in the education for emotionally and behaviourally disturbed children but also accentuates the possibilities for schools and units working with children who exhibit problematic behaviour. They show the regenerative nature of action research and its ability to promote not only research in this area but also ongoing support of staff and school development. They demonstrate how the adventure of acquiring language for self-exploration is introduced into cooperative play. Their conclusion is that the development of specific active involvement lends itself to the introduction of the relationship between behavioural and psycho-therapeutic approaches.

Part III begins with Anne Edwards' paper which asks if action research can give coherence to school-based learning in initial teacher training programmes. Anne argues that reflection on action is a vehicle for learning when combined with neo-Vygotskian dialogues around teaching and learning. She proposes a way of bringing reflection on practice into initial teacher training by connecting it to a framework for understanding how students may learn best in the settings in which they are placed for training. With mentoring processes linked to professional developments, school transformation may be possible.

Susan Cox's contribution complements that of Anne Edwards' in its focus upon alternative ways of thinking about how students learn how to teach. She writes about the theory/practice divide and the role of action within theories. It is an explication of action and practice in a reflective investigative framework based in school practice. She describes the FTS (Foundation Teaching Studies) course in the BEd award and how asking students to fill in critical incident sheets with the help of a critical friend, does or does not enable students to focus on the implicit theoretical constraints underlying their practice. Her paper provides ideas consonant with another contributor (Crowe) in her questioning of the relationship between feeling and action. She concludes that it is clear to her now how crucial the process of teacher education is to help students shift their conception of learning to teach, from a technical process of solving problems to a reflective process of problematizing action.

David Wilcockson writes his paper on pupil underachievement from a large project to establish records of pupils' educational, medical and social experiences. This paper begins to investigate the connection between poor educational, social, personal and medical histories and underachievement and the link between behaviour and underachievement. What makes this paper of particular interest to readers is the way in which teachers were drawn into the research. They began to realize that their actions affected the cycle of disaffection, deprivation and underachievement, and school targets for students were modified in response to emerging issues. The classroom support of teachers through team-teaching was seen to be helpful in generating and sharing resources and in supporting discipline. Through the action research, identified problems were addressed and faced within the school rather than being deflected outside to social/external factors as had happened in the past.

Rosemary Crowe investigates the way in which teaching and learning are experienced emotionally in a distance learning course. The course used for the focus of the enquiry is based on units of study that encourage critical reflection and the unit assignments are based on an action research model. She uses the students' writing as evidence of their feelings. These writings provide evaluative data for the assurance that the course experience was enlightening and empowering and offered insights into students' real experiences to consider difficult elements in the course design. The paper makes a plea for including feelings as a valid aspect of experiences that must be accounted for.

This book affirms the position of action research as a process for transformational practice in professional development contexts internationally. The discourse around student and teacher education validates the place of practical enquiry in teaching and learning and its potential for reconstructing stale and outmoded educational practices.

May I take this opportunity to thank all those named above for their contributions, who range from professors in education, through higher education lecturers and tutors, to classroom teachers. The range of their perceptions enriches the quality of this book and to them I am indebted.

Christine O'Hanlon
October 1995

References

SCHÖN, D.A. (1983) *The Reflective Practitioner: How Professionals Think in Action*, New York, Basic Books.

SCHÖN, D.A. (1987) *Educating the Reflective Practitioner: Towards a New Design for Teaching and Learning in the Professions*, San Francisco, Jossey Bass.

Part I

National Contexts in Action Research

1   Reflecting Where the Action Is: The Design of the Ford Teaching Project

John Elliott and Clem Adelman

In 1973, with a grant from the Ford Foundation, we began to set up an action research project into Inquiry/Discovery teaching with the cooperation of local authorities in East Anglia. What follows is an account of the project's origins and design.

Origins

The basis for the project's design originated in the work of the Schools Council Humanities Curriculum Project (HCP) under the direction of Lawrence Stenhouse (John Elliott was a member of its central team from 1967–1972). The HCP team faced the problem of supporting teachers who were experiencing great difficulty in implementing open discussion, aimed at developing adolescent students' understanding of controversial value issues. During the experimental period in schools from 1968 to 1970 many of its resources were directed towards helping teachers to adopt and test strategies aimed at resolving some of the most persistent and general problems they faced.

The Evaluation Unit of the Humanities Project, under the direction of Barry MacDonald, completed its measurement programme (Hamingson, 1972) and found that teachers who had received some training and support in adopting and testing the project's teaching strategies, tended to bring about results which were not paralleled among 'untrained HCP teachers' and 'non-HCP teachers' samples. These results — along such dimensions as reading, vocabulary, self-esteem, motivation, and concrete-abstract thinking — were such that teachers working in a variety of subject areas would find them desirable. They suggested four main implications:

1. That teaching strategies designed to give pupils greater independence from the teacher as a source of knowledge, and more autonomy over his/her own learning, (Inquiry/Discovery approaches), need not be responsible for a decline in educational standards as argued by the 'Black Paper'.

Problems that teachers in other curriculum areas had experienced in attempting to realize similar aspirations were investigated. The HCP team felt that it was not so much the aspirations themselves at fault as the distorted processes which often misrepresented their realization.

2.   It seemed that teachers trained to adopt the project strategies tended to promote generally desired, but largely unrealized, educational effects by implementing a classroom process which gave pupils greater independence and autonomy in the learning situation. This suggests that some of the strategies, at least, were control variables which generally prevented both the process and its effects from being realized. It pointed to the possibility of being able to identify problems and strategies which could be generalized across subject matters, classrooms and schools.

3.   The majority of the project's strategies rested on a diagnosis of classroom problems. These problems were connected with certain habitual and institutionally reinforced patterns of influence that teachers tended to exert on pupils. The project team found, for example, that teachers tended to foster dependence on the authority of persons rather than evidence and pressed for consensus conclusions rather than open discussion and an understanding of divergent views. Moreover, teachers were often unaware of the sorts of influence they tended to exert in classrooms.

Another implication of the measurement results showed that, since many of the problems faced by HCP teachers might be highly generalizable to other contexts, it would be reasonable to look in all contexts for a diagnosis in terms of general tendencies teachers have to exert certain types of influence on pupils.

4.   The training and support the HCP team offered to teachers was aimed at helping them to gain greater autonomy in their situation by systematically reflecting on their own actions and then modifying them in the light of an awareness of their consequences. The team encouraged self-awareness as a basis for decision making rather than dependence on themselves as authorities on classroom matters.

The measurement results might be seen to indicate that the pattern of inservice training and support which fostered systematic reflection on classroom practice has a practical pay-off in helping teachers to narrow the gap between aspirations and realization.

The Design of Support for Teachers

These four implications of HCP formed the basis of the Ford Teaching Project's design. We saw the tasks of the project and its central team in the following terms:

*To Help Teachers already Attempting to Implement Inquiry/
Discovery Methods, but Aware of a Gap between Attempt and
Achievement, to Narrow this Gap in their Situation*

We didn't try to involve teachers who made no claims to be attempting this
approach. Nor did we try to involve many self-styled 'experts' who felt they
had little to learn but much to offer. In the end we involved fifty teachers from
fourteen schools who seemed at the time to go some way towards meeting
these criteria. Inevitably we have since discovered some who were suffering
certain illusions about what it was they had been attempting. We felt these
numbers were the maximum that three central team personnel and three local
advisers (appointed by their Authority to take an active interest in the project's
work) could adequately support.

*To Help Teachers by Fostering an Action Research Orientation
towards Classroom Problems*

There are roughly two alternatives open to external agencies wishing to
make educational research relevant to particular classroom situations. One,
increasingly advocated, seeks to bring teachers and researchers together in a
cooperative enterprise. It involves different personnel with distinct but com-
plementary kinds of expertise. The teacher is the expert in classroom policy
making, but in order to function in this way he requires the systematic diagnosis of his situation, which only the researcher can supply. So the teacher
gives the researcher access to his problems and classroom. In return the researcher provides him with a diagnosis as a basis for decision making. This
conception of action research has its practical limitations. It can only have a
limited application in the absence of enough competent people in the field of
applied educational research to meet the likely demand for adequate research
support. Also, most action researchers have probably been trained in pure
research and are based in academic institutions. They are prone to the temp-
tation to sacrifice the practical requirements of action for 'academic standards'
and 'research purity'. Since, for teachers, understanding is necessarily instrumental to action, they require research support which is prepared to sacrifice
methodological purity to the needs of action. However, to date we have little
experience of developing research methods and instruments which furnish
teachers with quick but worthwhile diagnoses. The most fundamental problem
goes beyond the merely practical. The 'cooperative view' of action research
seems to us logically to imply dependence by the teacher on others for reflective analysis. Surely, this is inconsistent with placing great importance on the
teacher's power to perform his role autonomously and responsibly.

The degree of autonomy and responsibility a person exercises in his situation, the extent to which he is able to control the consequences of his actions, depends on three main conditions:

1. That he is aware of his own future actions; of what he will, or will not attempt to bring about. This awareness originates in a practical decision to do or attempt some things rather than others, and not in a prediction based on theoretical knowledge about his tendencies to act in certain ways in certain circumstances.
2. That a wide range of possibilities are open to him in a situation of which he is aware. Autonomy is restricted both by the extent to which choice is actually restricted and the extent to which his beliefs about his situation make him unaware of possibilities.
3. That he is aware of his potential relationship to his situation in the natural course of events; of his tendencies to influence his situation in certain ways. This is not to be equated with practical awareness (1) because it originates in detached reflection on the consequences of action rather than in a decision.

For the rest of this article we shall refer to these three conditions of autonomy as practical awareness, situational awareness, and self-awareness. Without spelling out for the moment the way these three forms of awareness relate to foster autonomy we would simply assert that the ability to act autonomously depends on the degree of awareness at each of these three levels.

The activity of teaching necessarily implies that the agent possesses the power to act autonomously. Inasmuch as it involves the attempt to bring about some kinds of learning rather than others, the first condition of practical awareness is present. One couldn't attempt X without having formed the intention to attempt X. Also, X couldn't be sincerely attempted if a person held no true beliefs about his situation and his own relation to it. Could a madman who seriously believed his pupils were stone carvings and himself to be their sculptor be seriously thought to be in a position to teach them? If the power to act autonomously is at least to some extent a necessary condition for teaching to take place then there is a sense in which a concern for a truer understanding of situation and self — which not improperly could be described as a research attitude — is a latent if not manifest aspect of the teacher's role.

The fundamental objection we have to the 'cooperative' view is the fragmentation it implies between practice and theory, and action and reflection on action, and the teacher's uncritical dependence on the researcher's expertise in the area of theory and reflection. This necessarily makes it impossible for the teachers concerned to develop the research function which is a latent and integral function of their role, thereby limiting their capacity to act autonomously.

In the Ford Teaching Project we opted for an alternative conception of action research where action, and reflection on action, are the joint responsibilities of the teacher.

To Support Classroom Action Research in the Area of Inquiry/ Discovery Methods

This support would be offered, firstly, by creating the beginnings of a shared tradition of thinking about teaching which would transcend such established educational 'frontiers' as subject divisions, classrooms, schools, and the primary and secondary sectors.

With this in mind we set corporate research tasks for our teachers which we hoped would set their situation-specific reflection within a wider context of mutual support. The tasks we set were as follows:

1. To specify the nature of Inquiry/Discovery teaching.
2. To identify and diagnose the problems of implementing Inquiry/Discovery methods, and to explore the extent to which they can be generalized within the project.
3. To decide on strategies aimed at resolving problems, test their effectiveness, and explore the extent to which they can be generalized.

We also established, and have a responsibility for maintaining and adapting an organizational framework which would facilitate the execution of these tasks. First, we got our teachers to agree to meet regularly in school-based teams to compare and contrast experience. Then, we asked headteachers to appoint one member as team coordinator, responsible for convening meetings, liaison with us, and supporting the team's work in the school generally. Third, we arranged for school teams within easy reach of each other to meet regularly at a nearby Teachers' Centre. These meetings were convened by a local adviser and we did not attend unless invited. This enabled teachers to criticize our work with them freely and the adviser to report problems which may not have been aired in our presence. Fourth, we arranged for all the teachers involved to meet three times during the two years of the project at 3–5 day residential workshops. The first 'launching' conference was held during the Easter vacation.

The tasks set and the support framework outlined are based on a certain assumption which we believe it is our responsibility to test: This is the assumption that it is possible for a group of teachers working in different contexts to produce generalizations about problems, explanations and strategies.

A second method of supporting action research would be by helping teachers to be more aware of the consequences of their actions by systematically monitoring pupils' accounts of them (Adelman and Elliott, 1973). Accurate self-monitoring by teachers is an important aspect of classroom action research for two reasons. First, teachers need to know whether or not the strategies they decide to adopt produce the consequences intended. If they are ignorant in this respect they are not in a position to identify and resolve obstacles in their path. Secondly, we assume that many of the most persistent and generalizable problems in the area of Inquiry/Discovery teaching can be diagnosed in terms of teachers' tendencies to influence their situations in ways of which they are largely unaware. In spite of what is being attempted a teacher's actual performance may bring about unintended effects which ensure failure. If he is to be in a position to resolve problems created by his own actions, he must become aware of things he brings about unintentionally.

One might ask what systematically monitoring pupils' accounts of the teacher's actions has got to do with his becoming more self-aware. Let's examine more closely how an agent identifies another's response as a consequence of his actions. This isn't simply a matter of correlating an action with some event which follows it. An event has to be identified as something brought about by his agency. In other words, he must know 'the causal mechanism', to quote Harre and Secord (1972), which relates his action to its consequence. One might think that with intentional action this is a simple matter, an event which corresponds to the teacher's intentions or ends-in-view surely provides a criterion for identifying it as something he brings about. However, an agent can attempt to bring some consequence about and the event intended may subsequently occur without this being either necessary or sufficient to identify it as something that s/he brings about. So we must search for the causal mechanism which relates acts to their consequences — intended as well as unintended — in a different sphere to the agent's intentions. And here we have to turn to the meanings ascribed to actions by those on the receiving end.

In a human situation, consequences of actions are identified as such, not so much in terms of the agent's intentions, as by the way other people interpret them. It is other people's interpretations of an action's meaning that in the final analysis explain the relationship between act and consequence, and provide criteria for identifying an event as a consequence of one's actions. If it can be shown that a particular event is a response to some action because it is based on an intelligible interpretation of its meaning, then the agent can be said to be responsible for what happened. A consequence that is intentionally brought about is one where the agent's and recipient's interpretations are agreed. It expresses mutual understanding of the significance of actions. A consequence that is unintentionally brought about may not simply express someone's refusal to respond in the way intended, but a discrepancy between the agent's interpretation of his action and the recipient's.

If this account is correct then people can only become aware of what they actually bring about in their situation by understanding how their actions are interpreted by others. In normal circumstances the production of certain kinds of response will be reasonable evidence that actions have been interpreted in the appropriate light. This is made possible by the existence of shared rules governing how actions are to be interpreted in social situations. For example, in classrooms certain kinds of teacher behaviour are characteristically viewed by pupils as expressing certain intentions. These interpretations are governed by rules which have been acquired from, and reinforced by, their contact with teachers over the years. The rules lay down what teachers can be expected to mean when they do certain things. When the rules are shared, say by teachers and pupils alike, there is little point in eliciting accounts from others about how one's actions are interpreted. But there are situations in which people cannot assume shared rules. If an agent is to get himself in a position here accurately to monitor the consequences of his actions, he must systematically elicit *other's* accounts of what he is doing.

Innovations in teaching, such as attempts to implement Inquiry/Discovery methods, are necessarily exploratory situations where consensus on rules cannot be assumed. Here teachers are trying to bring about new possibilities in contexts where there are firmly established and time-honoured rules governing the interpretations of their actions, setting limits on what they can possibly be expected to intend and consequently on how pupils can be expected to respond. Effective innovation depends on teachers and pupils negotiating new rules, and is stifled by pupil responses based on misinterpretations of a teacher's behaviour in terms of established rules. Here teachers need to be constantly discussing their own accounts of their conduct with those of their pupils if they are to close the gap between attempt and achievement.

At the present time systematic self-monitoring along these lines is largely undeveloped in classrooms where teachers are trying to realize new possibilities. The result all too often is that teachers fail to identify some of the most persistent problems they face as unintended consequences of their own actions. The problems persist because they remain unaware of their relationship to them, and they remain so because they receive little institutional support *to* develop self-monitoring methods appropriate to innovation. When searching for teachers and schools to join the Ford Project, for example, we found large numbers of 'innovators' being kept so busy in their ever-extending institutional commitments that they couldn't spare the time to do much research into their own teaching.

To Ensure that Teachers' Action Research is Carried Out in a Way which Protects and Fosters Autonomy

We believe we have a responsibility to design the action research methodology in a way that ensures a creative interaction between action and reflection. The design is based on the following philosophical assumptions:

- *That prespecification of what is to be attempted necessarily sets limits on the development of autonomy if teachers take it as a guide to action*

There is an important distinction between accurately describing and being aware of what is being attempted in Inquiry/Discovery teaching. Hampshire (1965) writes:

> I may be trying to do something, and going forward with a fixed and definite intention in my mind, and still be liable to make a mistake of some kind in characterising my action in words. (Ch 2, p. 961)

As we suggested above, awareness of what is being attempted is practical, originating in the formation of an intention. In some cases this may be an intention to bring something about under a particular description. Here a person's awareness of the object of his intention presupposes a prior description of that object. However, there are other cases where the relation between intention and object is not mediated by a description; it is direct. Here one may be quite aware in one's own mind of what is being attempted but misdescribe it.

The problem with the first kind of case is that it places certain logical limits on the subsequent development of practical awareness. If teachers pre-specify at the beginning of the research what is to count as Inquiry/Discovery teaching and then decide for the rest of the project only to intend those things falling under it, they restrict the practical thought expressed in their subsequent attempts to the degree of awareness expressed in the prespecification.

Intentions can be more or less vague and precise, and there is always some blur at their edges which can be refined. Hampshire (1965) argues that 'No sense can be given to the idea of an absolutely specific intention . . .' (Ch 2, p. 123) We believe that any corporate specification of Inquiry/Discovery teaching by our teachers should genuinely reflect a corporate development towards greater practical awareness of its nature. Consequently we feel it is our responsibility to resist pressures from our teachers (already apparent) either to get us to prespecify or to try to reach agreement themselves on exactly what is to be attempted during the programme.

We are not objecting here to specification or even prespecification when it is simply used by teachers to communicate their practical awareness to each other. Here it is used as a description of intention formed prior to it. All we are trying to do is to ensure that the requirements of rational action are not confused with the requirements of communication. Even if at the end of the project, our teachers are able to produce some corporate specification we hope that this will serve to communicate their corporate awareness to others rather than restrict further action research in the area.

- *That people clearly identify practical problems in the light of their intentions rather than a prespecification of intention*

Teachers may disagree about *what it is* they are attempting but agree in what they attempt and vice versa. It follows from this that an agreed prespecification is not a necessary condition of being able to identify shared classroom problems (as some of our teachers have again argued). We adopted the view that one identifies shared problems by first identifying a problem in one's own situation. And here a prespecification is not necessary; only a degree of practical awareness of the objects of intention. Intentions are, to quote Hampshire again, '. . . like a torch throwing its light forward, illuminating, resisting objects in their path.' It is by virtue of some degree of shared practical awareness that teachers can come to identify and talk about shared practical problems.

- *That as practical awareness develops new problems are identified for action research*

We see action research which protects autonomy to be a process that is dynamic rather than static. As practical awareness develops new problems constantly emerge for diagnosis and decision. The use of prespecification, we have argued, destroys this process, restricting the range of problems which can appropriately emerge, and consequently the range of strategies tested. It expresses a 'stop the world we want to get off' attitude. We believe we have a responsibility to ensure that our teachers do not sacrifice autonomy on the altar of security.

- *That people can discover the extent of their agreement on what is being attempted by identifying shared practical problems rather than trying to describe their intentions*

At our initial conference we argued that people would have evidence that they were attempting similar things in the classroom if during the research they identified similar practical problems. In one group, during an initial exploration of the problems teachers had already encountered, some became disturbed by the sorts of problems others had raised; they didn't fit their conception of Inquiry/Discovery teaching. They argued that the only way to ensure agreement on problems was to agree at the conference on a prespecification of what was to be attempted. We were very sceptical about the possibility, let alone the desirability, to such a consensus, since we believed that any agreement on an accurate and precise description would presuppose a level of agreed practical awareness which was unlikely to exist. Having tried and largely failed to achieve a precise consensus specification some doubted whether there was any basis at all for future cooperative research with each other. We felt that such doubts failed to take into account the possibility that as the project proceeded, individuals would modify their practical awareness of what was being attempted in a way which brought them more together. Evidence for this would take the form of being able to identify shared problems. In the meantime we hoped that our teachers would respect differences of view and not seek a false security in attempts to pressurize each other into an agreed prespecification which, if successful, would only in the long run stifle the autonomy of the individuals involved. We see it as our responsibility to ensure that the autonomy of individuals is not sacrificed by a desire for consensus, and that consensus in practical awareness develops within a context where practical thought is not restricted.

- *That practical awareness not only determines the requirements of action research but is in turn modified by its products*

So far we have outlined ways in which action research can be designed to protect autonomy and the growth of practical awareness and how it is influenced by this growth. But we believe that it can also foster autonomy by influencing practical awareness. In other words, there is an interaction or reciprocal relation between action and reflection. This is because self-awareness of one's tendencies to bring about certain things in a situation can modify one's beliefs about what is possible in it (situational awareness), which in turn may bring changes in practical awareness.

An example, drawn from the Ford Project, may suffice to illustrate these relations. At our Easter Conference some of our teachers argued that the aims of Inquiry/Discovery teaching are no different from those of more traditional approaches:

> I still hold the opinion that the ends are the same, where you want the child to finish up is much the same as traditional authoritarian teaching with chalk and talk.

They argued that the difference lay in the way the teacher proceeded to achieve these common ends. In traditional teaching he told pupils what to believe. In Inquiry/Discovery teaching he lets them reason out the conclusions independently:

> . . . in coming to this [conclusion] he won't take it just as a fact from myself — Sir told me so and take it on that strength alone — but because he reasoned it out himself.

This conception of Inquiry/Discovery teaching rests on the belief that it is possible both to predetermine the learning outcomes and to foster the autonomous reasoning of the pupils at the same time. In other words, it rests on certain beliefs about what is possible to achieve in one's situation. If a teacher believed that one could not possibly foster autonomous reasoning while predetermining the learning outcomes then he could not sincerely attempt both at the same time. As Hampshire (1965) argues, a person's will and intentions cannot be determined independently of the beliefs (s)he holds about the situation they confront (see Ch 1, pp. 84–85).

During the first term of the Ford Project, schools have evidence that some teachers who viewed Inquiry/Discovery teaching in the way outlined have shifted. They still view it as an attempt to foster autonomous reasoning but not as trying to predetermine the learning outcome. So there is both continuity with, as well as modification in past awareness of what is being attempted. These modifications rest on changes in situational awareness of possibilities in their situation. They no longer believe that one can both predetermine learning outcomes and foster autonomous reasoning at the same time in their situation. These changes in turn were made possible by the more systematic self-monitoring the teachers were now engaged in. They discovered that in pursuing certain learning outcomes that fostered dependence rather than autonomy, and this awareness called into question their beliefs about possibilities in their situation, and consequences of what could be intended in it under the label of Inquiry Discovery teaching.

To Test the Following Hypotheses

1. That it is possible for a group of teachers working in a variety of contexts to identify problems and effective strategies for resolving them which are highly generalizable.
2. That action research methods which promote self-awareness by monitoring pupils' accounts of teaching are the best means of helping teachers fruitfully to diagnose their most persistent and generalizable Inquiry/Discovery problems.

We have already made explicit two assumptions implied by the ways we are trying to support action research in classrooms. These are reformulated above as two hypotheses which we believe we have a responsibility to test. So, it might be argued, we do see ourselves as researchers after all. This is so. However, we see our research tasks as somewhat different to those of our teachers, although logically related to them. Whereas our teachers are doing action research into Inquiry/Discovery teaching we are doing action research into effective ways of supporting action research of this kind. Our research task on the central team must be seen in relation to the aims of the project and not the teachers involved in it. This important distinction is often blurred.

The hypotheses the project itself are testing are 'theories about theories'. They are about the possibility of getting teachers to produce worthwhile generalizations across their own experience and the possibility of testing them to explain their most persistent problems in terms of their own unconscious influences on their situation. In other words they are second-order theories about the possibility of a first-order theory of Inquiry/Discovery teaching, and are tested by the ability of our teachers to produce the latter. Our action research is therefore logically dependent on that of our teachers.

We are making the assumptions behind the action research tasks, framework, and methods adopted by our teachers, explicit to them in order to show we are prepared to test and modify them in the light of their efforts. If we did not do this, we would hinder their capacity to reflect and encourage an uncritical dependence. Any educational support agency which claims to value the teacher's power for autonomy in the classroom can be accused of proceeding irrationally if it fails to make its assumptions explicit and to invite teachers consciously to participate in the task of critically testing them.

References

ADELMAN, C. and ELLIOTT, J. (1973) 'Teachers as evaluators', *The/ ma Curriculum* (Germany), Nov. 1973 and articles by the Ford Project Team and teachers in *The New Era*, Dec. 1973.

Hamingson, D. (1972) (ed.) *Towards Judgement,* the publications of the Evaluation Unit Of the Humanities Curriculum Project 1970–1972, Ch VI Part 2 CARE Occasional Publications no. 1, University of East Anglia.

Harre, R. and Secord, R. (1972) *The Explanation of Social Behaviour,* Oxford, Blackwell.

Hampshire, S. (1965) *Thought and Action,* London, Chatto and Windus.

2   Repositioning the Teacher in US Schools and Society: Feminist Readings of Action Research

Sandra Hollingsworth

This chapter addresses the intersection of feminist analyses of lived educational experience and the action research movement. In hopes of establishing a conversational tone throughout the text, I will begin by telling you — the reader — a partial story of my life's education and work so that you will come to know me as a human being who may share some of your own lessons and struggles. I tell the story as a way of inviting you to engage in a project of relational, as well as critical, understanding of the text. I'll then turn to the social and political climate which gave meaning to my life's work through feminist theory, describe various feminisms, and situate the importance — for me — of merging feminist theory and action research. I won't address action research in detail here, however, as I have done so in two recent publications (Hollingsworth, 1994b; Hollingsworth and Sockett, 1994).

An Autobiographical Backgrounding to my Current Position

I grew up internationally in a working-class family — my father's avocation was to experience life as a 'swing era' musician and my mother's was to care for our home. Yet a significant part of my education took place in the southern United States, after my father's untimely death at the age of 37, during a period of racially segregated and patriarchal schooling. My interests in social justice came through my own education. As editor of the white high school newspaper, I finally won the 'freedom' to write about the inequities between the white high school band, housed in a brand new air-conditioned brick building — and the black high school band, practicing in a windowless structure on a dirt road, with damaged instruments and colorless uniforms that we had passed on to the black high school after twenty years of use. I never wrote about the inequities in my own education, however, until much later. Only then did I see the significance of my being taught 'world' history from the perspective of only white, European-American men, or my being refused entry to advanced maths classes because 'girls didn't need that level of mathematics.'

In the face of these obstacles (and encouraged both by a grandmother and a high school teacher who believed in my questions), I went to college and — now open to the fuller stories of world history — became an historian. Later, when my own son had difficulties in school, I returned for two more degrees in education, and became a classroom teacher — first at the secondary level, and then at the elementary level. Although I became better able to understand some issues of curriculum and instruction, I still could not help my son negotiate the political structures of schooling. So I returned for a doctorate.

My education in the doctoral program at the University of Texas, Austin, centered on 'training' me to become a researcher in the 'apolitical' tradition of educational psychology. My own questions about educational practice went outside that disciplinary boundary, so I also educated myself in anthropology and sociology, but even my best arguments to 'see' the knowledge of education in other than cognitive ways were judged as wanting. Again, however, because a teacher, Walter Doyle, believed in me, I took my first postgraduate post as an Assistant Professor at the University of California, Berkeley.

Following the mode of my own schooling, now positioned as a teacher educator, I attempted to 'transfer' what I had learned about education and educational research to teachers-to-be. I failed miserably. How did I know the extent of my failure? I researched my own teaching, following the teacher candidates I taught into their first seven years of practice. Although, under the coercion of my evaluative authority in university courses, they had appeared to learn what I knew, they really needed to know many other things about teaching and learning in urban school settings that I had never dreamed of. They needed the political knowledge that I also needed (but didn't get through my graduate program) to help my son through school. They needed to engage in the praxis of learning to teach — or understand the world of education in order to change it.

So I turned to the action research literature as a guide, as it was the closest approximation to educational praxis I knew then. Influenced by various conceptions of action research, such as Lawrence Stenhouse's (1975) conception of teachers as researchers; John Elliott's (1991) notion of action research as a pedagogical paradigm or form of teaching; descriptions of the task as a 'systematic self-reflective scientific inquiry by practitioners to improve practice' (James McKernan, 1991); structural/institutional critiques of institutions as workplaces for teachers (Carr and Kemmis, 1986); or a philosophical stance toward 'the inter-dependency of knowledge and action, of educational theory and educational practice' (Somekh, 1993), I came to believe that cooperating with groups of teachers engaged in action research was the way to facilitate their learning to teach. I also taught my own courses at Berkeley as praxis (see Hollingsworth, 1994a).

Yet even action research, situated apolitically at the intersection of knowledge and practice, was not enough. Positioned in the lower levels of a hierarchy of educational responsibility where abstracted knowledge of teaching is more valued than useful knowledge (see Labaree, 1993), the teachers I worked with at Berkeley did not perceive themselves to be powerful enough to 'create and critique knowledge' from external authorities. Even while they realized that external and generic research 'findings' — abstracted from their specific persons and practices — were insufficient to successfully teach their urban students, they did not experience the *agency* to construct alternative knowledges. I wondered why. My search into the literature this time led me to feminist theories of the social position of teaching.

Feminist Theories and Educational Practice

Feminist literatures resonated with my experiences as a classroom teacher and teacher educator. I began to see clearly that teachers, as a class, work under less than professional conditions with increasingly complex demands for academic, social and psychological expertise in demographically diverse settings. Yet teachers are asked to comply with a narrowly constructed set of ideological standards which shape and evaluate their practices. These standards — stated as credentialing requirements, teacher and student evaluation measurements, curricular mandates, and even appropriate research paradigms for advanced degree work — are historically established by people outside of the classroom in positions of power (mainly men) without benefit of teachers' (mainly women) voices and opinions (see Apple, 1985). Further, teachers are educated in institutions of higher education where paradigms other than those traditionally and problematically associated with masculinization are devalued (Martin, 1981, p. 104): 'The "male epistemological stance" becomes everybody's stance' (Brittan and Maynard, 1984, p. 204).

The result of this political and social assignment places both male and female teachers (as a class) in the position of 'Other' where 'one's existence is necessarily outside of and apart from the public flow of discourse and meaning' (de Beauvoir, 1949, pp. xvi–xxix) and calls for feminist critique. Glorianne Leck (1987) tells us:

> Of central importance in the feminist critique is the examination of the primary assumption of patriarchy — that activities of male persons are of a higher value than activities of female persons. This precept is woven into the entire intellectual paradigm that is foundational to current schooling practice. For this reason the feminist critique orders a challenge to the epistemological roots of educational theory as we know it. (p. 343)

Upon reflection of my experiences with teachers engaged in action research, I became awake to the different positionings of 'teacher' in this work. That is, both university instructors who engage in research on their own practice and school-based teacher-researchers regularly question their teaching and their students' learning; collect information to inform themselves about those questions; experiment; document; summarize; and try again. The distinguishing aspects of classroom teachers' reflective work from more traditional research reflections occur in the areas of documentation, report and influence. Unlike enquiry performed and reported by university scholars where time, reward, and authority are considered part of their jobs, school-based teachers' reflective processes are rarely given sufficient time to develop critically, rewarded, or observable to others. Within teachers' daily lives, in fact, they may perceive neither time nor reward for articulating the processes — even to themselves. As a result, they (and others) undervalue their natural ontological capacities for reflectively learning from experience. Both the undervaluing of their work and the conditions that mitigate against reflection are unfortunate characteristics of American public school teachings — or the institutionalization of 'women's true profession' (Laird, 1988).[1]

There are, of course, many interpretations other than feminist to explain the low status of teaching in the US, a number of which reflect the economic or marketplace ideology which drives our society (Bellah, Madsen, Sullivan, Swidler, and Tipton, 1987). David Labaree (1993), for example, points to market pressures of teaching and teacher education, which give them lower social status. Labaree's argument certainly has face validity. Yet he fails to clarify how gender is another aspect of market. He does tell us that the common school movement in the nineteenth century, for example, created a market for 'cheap' teachers. What he might have also said was that two women could be hired as teachers for the cost of one man (Clifford, 1989). When the first wave of feminism awakened women to the fact that they could now claim an education, even elementary teachers in 'normal' training schools demanded more than the professional agenda they were offered, and eventually moved into the universities — whereupon teacher educators began to share the low status which accompanied their gendered association with teachers whose work is closest to children (Warren, 1989).

Leaving gender out of discussions of the status of professions does not help us understand why other professionals, such as doctors (mostly male), enjoy a higher status because of the exchange market value of the medical field, while nurses (mostly female) do not share the same privileges within the same field — nor why useful knowledge (often more applicable to women's occupations) is less valued than abstract knowledge (available to valued professionals). The gendered neutrality of a market analysis in any profession appears insufficient.

Further, because of the supposedly 'neutral' position of the profession encompassing schooling, pedagogy and psychological research (and the lack of contrary canons within our educations to tell us otherwise), it is difficult to see the relations of power in education. Australian educators Jane Kenway and Helen Modra (1992) tell us that discussions of pedagogy (a term used more by academics and less by classroom teachers) often rest

> . . . upon an instrumental, transmission model of teaching which fails
> to make problematic either the learner, teacher, or knowledge, or the

relationship among them. Certainly the common conception of pedagogy is blind to the ways in which broader social relationships are embodied in the teaching/learning process. (p. 140)

Feminist readings of teaching, learning and educational research render the social and political relationships across them problematic — in different ways. There are, perhaps, as many conceptions of feminism as there are feminists. In fact, US literacy scholar Carolyn Burke (1978) claims that the strength of the women's movement lies in its ability to acknowledge serious disagreement among feminist positions. It is not my intention here either to illuminate them all or to categorize them for easy handling. Therefore, I do not want to classify feminisms either chronologically in terms of first, second and third waves.[2] Nor do I want to characterize them according to 'traditional' conceptions of liberal, radical, socialist/Marxist, black, lesbian or age-conscious feminisms, which seem to isolate each position from the other unnecessarily (see Arnot and Weiler, 1993). However, I do want to point out the ranges of feminist theory and research so that those unfamiliar with the field can get a sense of its diversity and self-critical nature. I will, therefore borrow from the stances Michelle Fine uses (1992) to situate feminist psychologists in the texts they produce: ventriloquy, voices, and activism.

Ventriloquy

Beginning with a critique of mainstream psychology which flattens, neutralizes and objectifies personal and political influences upon its 'discoveries of natural laws of human behavior,' Michelle Fine (1992) uncloaks researchers who 'pronounce truths while whiting out their own authority, so as to be unlocatable and irresponsible' (p. 214). She reminds us of Donna Haraway's (1988) caricature of this epistemological fetish with detachment as a 'God trick . . . that mode of seeing that pretends to offer a vision that is from everywhere and nowhere, equally and fully' (p. 584). She quotes for us Lilian Robinson:

> Once upon a time, the introduction of writings of women and people of color were called politicizing the curriculum. Only *we* had politics (and its nasty little mate, ideology), whereas *they* had standards (Robinson, 1989, p. 76) . . . Ventriloquy relies upon Haraway's God trick. The author tells Truth, has no gender, race, class, or stance. A condition of truth telling is anonymity, and so ventriloquy. (Fine, 1992, p. 212)

At the beginning of my career in higher education, I used this methodological stance — measuring teachers' knowledge against a backdrop of cognitive and social theory. I never directly asked the teacher 'subjects' who worked with me to comment on my evaluation of their minds' and their bodies' work. They remained hidden in my text, ostensibly to protect *their* identities — but, just as well, to protect me from their reading and rejecting my interpretation (see Hollingsworth, 1989).

Voices

Becoming aware of the blanket silencing which comes with ventriloquy, many feminist researchers (including me) began to reject its 'neutral' stance in research, and to articulate and amplify 'feminine' voices. Beginning with Carol Gilligan's moral critique of 'human' development *In a Different Voice* (1982), Nell Noddings' ethic of care (1986, 1992), and Sandra Harding's epistemological critique of knowledge (1986), each of whom, in turn, drew upon such feminist object relations theorists as Nancy Chodorow (1978), these feminists have engaged in theorizing gendered subjectivity from a 'difference' perspective. They argue that the socialized experiences of girls and women teach them to view identity, morality, and even education differently from boys and men. In fact, it has been my experience (and the experience of many other feminist teachers and teacher educators) that teachers initially become aware of feminist critiques of knowledge and education through an awareness of difference. One of Magda Lewis' students reports:

> In history we never talked about what women did; in geography it was always what was important to men. The same in our English class, we hardly ever studied women authors. I won't even talk about math and science . . . I always felt that I didn't belong . . . Sometimes the boys would make jokes about girls doing science experiments. They always thought they were going to do it better and made me really nervous. Sometimes I didn't even try to do an experiment because I knew they would laugh if I got it wrong. Now I just deaden myself against it, so I don't hear it anymore. But I feel really alienated. My experience now is one of total silence. Sometimes I even wish I didn't know what I know. (Lewis, 1992, p. 173)

Feminist research in the 'voices' genre also led to classroom-based research in the United States which illuminated the lack of 'gender equity' or unequal classroom and curricular opportunities for girls.

> . . . the results of gender and education research filtered through to state and federal commission and inequities into girls and schooling, and subsequently into educational policies and curricula . . . Equal classroom time, equal numerical participation, and equal curricular presence were the main aims . . . (Luke and Gore, 1992, p. 8) (see also Klein, *et al.*, in press, and the American Association of University Women's Report *How Schools Shortchange Girls*, Wellesley, 1992 for similar issues)

Of course, what the early difference-feminists neglected to point out is that girls' and women's experiences in school differ by race and class, as well as gender (and other positionings — see Biklin and Pollard, 1993). bell hooks (1986, 1990) and Audre Lorde (1984), for example, remind us of the different positionings that black lesbian feminists occupy compared to white hetero-sexual females. Angela Davis (1981) reminds us how the nineteenth-century women's movement in the US began by association with anti-slavery cam-paigns — and thus owe a great debt to black women's unique experience. Many feminists have been theoretically critical of essential 'difference' femin-ism in all of its guises, including feminist pedagogy.

> . . . while males have a psychic preference for autonomy, separation, certainty, control and abstraction, females are differently connected to the world through their relational and contextual preferences and their superior capacity to offer empathy and tolerate ambiguity. This type of thinking lies behind the belief that 'girl friendly' pedagogy should emphasize interactive, cooperative, intuitive and holistic ways of learning. In certain senses, this argument sits comfortably with the beliefs of many humanist and progressive teachers that all students benefit from modes of pedagogy which harness affect and encourage group learning. However, from a[nother] feminist point of view, it has some problems. Too often in the feminist pedagogy which arises thus, definitions of girls' ways of learning and of girls' interests and motivations are developed which tap right back into the gender stere-otypes from which escape is sought. (Kenway and Modra, 1992, p. 145)

Philosophers such as Maxine Greene take still other approaches to difference feminist pedagogy. Greene (1978) reminds teachers in feminist classrooms to identify and speak for a wide-awakeness to the moral and aesthetic dimen-sions of existence.

> The problem, most will agree, is not to tell [students] what to do — but to help them attain some kind of clarity about how to choose, how to decide what to do. And this involves teachers directly, imme-diately — teachers as persons able to present themselves as critical thinkers willing to disclose their own principles and their own reasons as well as authentic persons living in the world, persons who are concerned — who care. (pp. 47–8)

Closely related to philosophical reminders of the importance of 'knowing and naming our current selves,' is Mary Field Belenky, Blythe McVicker Clinchy, Nancy Rule Goldberger, and Jill Mattuck Tarule's (1986) work on feminist epistemologies. For all of the omissions that the work failed to describe — particularly around race and class — it awakened me (and many others) to the importance of claiming personal theory and experience as valid ways of knowing. The authors demonstrated how some women, secure in their own interpretation of reality, create their own knowledges as well as critically examine others'. Those who were most distant from personal knowledge interpretations had to rely solely on others' objective observations as truth.

Other educational feminists turned their attention away from the perspective of male and female differences or philosophical constructions of lived experience and toward the discourse of critical pedagogy, poststructuralist and postmodern educational theory (following Cherryholmes, 1988; Freire, 1970; Foucault, 1980; Giroux, 1988). Liz Ellsworth (1989), Jennifer Gore (1992), Patti Lather (1991), Mimi Orner (1992), Valerie Walkerdine (1986) and others embraced the construction of 'critical feminist pedagogies' to 'liberate' the voices of the 'Other.' However, they found the application of critical theory to classrooms problematic. Like the object relations perspective of the 'girl friendly' pedagogy noted above, 'critical' feminist pedagogy still seemed trapped within modernist enlightenment epistemologies. The critical researcher, though now 'coming out' politically, still remained invisible with respect to his or her subjectivity in relation to the 'Other.' Thus, efforts to create 'emancipatory classrooms' within a unitary or transcendental (if critical) view of emancipation failed.

I've noted similar problems. One teacher, married to a multinational corporate vice-president told me after one class that she just didn't 'get' the feminist message. 'I went home and told my husband, "I've tried and I've tried, but I just don't feel oppressed!"' bell hooks and other feminist writers give me courage to continually try constructing critical classrooms. 'Education for critical consciousness is the most important task before us,' she writes. 'There are many individuals with race, gender and class privilege who are longing to see the kind of revolutionary change that will end domination and oppression even though their lives would be completely and utterly transformed . . . the feeling of "yearning" opens up the possibility of common ground where all these differences might meet and engage one another' (hooks, 1990, pp. 5, 13).

The best way I've found for dealing with this dilemma is to begin classes autobiographically, so that our positions on feminist theory can be situated; then explain how and why I see critical feminist pedagogy as a means to uncover and critique our own and others' theories; make the course evaluation problematic; and let go of class outcomes. I have to keep reminding myself that whether students adopt my current stance on feminist theory or not is NOT the goal of feminist pedagogy — it is to open us all up (even me) to new ways of seeing the familiar and challenging injustice.

Carmen Luke and Jennifer Gore (1992) help me remember the differences between feminist and critical pedagogy. They describe how situated and critical knowledges are often contradictory. Even as feminists learn from poststructuralist sociologies, our situated positions allow us to critique them — and those we reconstruct. I've learned much from humbling self-critiques of my own work (see Hollingsworth, 1994c). This stance, rather than hindering feminist projects, seems to provide a space to dance between modernist and postmodernist critical projects. At the very least, the awareness that there *are* differences in lived stances toward critical theories keeps us honest. Let me give some examples.

Inside and outside of the classroom, the attempt to 'empower' the 'Other' into a theoretical framing coming from the researchers' position of authority often led not only to 'voice' but to voicing a critique of the researchers' position. I've often convinced my students that it is appropriate to critique authority, and then feel slightly confused when they challenge *me*! One full class of pre-service teachers at Berkeley was ready to mutiny over my suggesting that they construct their own knowledges too fast. I can (I think even thankfully) say I've now had enough experience with this phenomenon that I now embrace it. However, sometimes the outcomes are less embraceable. Sometimes 'empowered' students advocate interventions and strategies that we organizers can't support (Fine, 1992, p. 217). One group of teachers I worked with in an urban school wanted to eliminate lower-track failure by barring entry to students from single parent homes! I eventually left that 'collaboration.'

Whether substantiating 'our' notion of empowerment or not, a romantic reliance on the 'voices' of the 'Other' in research — 'as though they were rarefied, innocent words of critique — represents a sophisticated form of ventriloquy, with lots of manipulation required' (Fine, 1992, p. 216). Shulamitz Reinharz (1992) reminds us that to hear another's voice, we have to be willing to hear what someone is saying, even when it violates our expectations or threatens our interests: 'In other words, if you want someone to tell it like it is, you have to hear it like it is' (p. 16). Many times, feminist projects of praxis which aim to illuminate the voices of 'oppressed' women, want to empower them, as well as, represent their lived conditions, while the 'oppressed Other' simply wants to join in their emancipators' critiques of schooling. 'With all of their suspicions of public education (and they expressed many), they still saw public schooling as the only possible vehicle for their children's futures. For many, the risks of voicing critique were simply seen as too great' (Fine, 1992, p. 217).

One of my greatest lessons in this work is to learn that others may not be emotionally, intellectually, or occupationally ready to take on the risks that I do. I need to constantly keep in mind that I am positioned as a tenured faculty member, a single mother with grown children, an economically stable head-of-household with a strong network of friends. My risks are not as costly as others. Further, it is often difficult to remember that an ultimate goal of 'voices' research is to establish conditions wherein the 'Other' can represent themselves, and not be represented by advocates. The teachers with whom I work, for example, remind me over and over again that our writing, publishing, speaking and challenging in academic circles is NOT the medium *they* need to be heard in their schools.

Additional critiques of what constitutes the 'voiced' nature of pedagogy include cautions against 'non-hierarchical' practices. Janice Raymond's book on female friendships (1986) suggests that such practices can prevent women from discovering and using their own strength, and also encourage them to endeavor to achieve their goals through the exercise of indirect power or even manipulation within a group: 'No real power emerges from a group that silences its best and brightest voices for a false sense of group equality. And certainly no strong friendships can be formed among women who have no power of being' (p. 197).

Finally, the critiques raised in assessing the 'voices' category of feminist research argue against 'swimming in the murky waters of essentialism' (Stanley, 1990, p. 14). Liz Stanley continues (with my emphasis):

> I am referring to a specifically feminist ontology, not an ontology attached to the category '(all) women.' *I make no claims that 'women' will share this state of being; patently, most do not.* . . . That it is the experience of and acting against perceived oppression that gives rise to a distinctive feminist ontology; and it is the analytic exploration of the parameters of this in the research process that gives expression to a distinctive feminist ontology . . . My concern is with the conditions under which some classes of people, but not others, are treated as, or come to feel they are treated as, 'other'; and consequently construct . . . a distinctively sociological epistemology. There is also nothing about the acknowledgment of 'difference' that precludes discussion, debate and a mutual learning process. (pp. 14–15. See also Britzman, 1993, for another excellent critique of gendered essence.)

Yet, as Fine (1992) reminds us,

> this critique of voices is by no means advanced to deny the legitimacy of rich interview material or other forms of qualitative data. To the contrary, it is meant for us to worry collectively that when voices — as isolated and innocent moments of experience — organize our research texts there is a subtle slide toward romantic, uncritical, and uneven handling and a stable refusal by researchers to explicate our own stances and relations to these voices. (p. 219)

Activist Feminist Research

A third stance toward theories and projects of feminist research takes into consideration the critiques of both the ventriloquy and voices research, and thus raises new questions. How can we position ourselves as researchers committed to critical, self-conscious and participatory work dedicated to change, engaged with but still respectfully separate from those with whom we collaborate? *How can we not only know differently from how we were educated, but continue to know differently?* For Michelle Fine (1992) and Patti Lather (1986), images of activist feminist scholarship share three distinctions:

First, the author is explicit about the space in which she stands politically and theoretically — even as her stances are multiple, shifting, and mobile. Second, the text displays critical analyses of current social arrangements and their ideological frames. And, third, the narrative reveals and invents disruptive images of what could be. (Fine, 1992, p. 221)

In activist research, the power of hegemony becomes fragile. Alison Jaggar reminds us of Antonio Gramsci's (1971) notion of hegemony as a concept 'designed to explain how a dominant class maintains control by projecting its own particular way of seeing social reality so successfully that its view is accepted as common sense and part of the natural order by those who in fact are subordinated to it' (Jaggar, 1983, p. 151).

Feminist activist research consciously seeks to break up social silences to make spaces for fracturing the very ideologies that justify power inequities — *even feminist ideologies.* It takes on a form of praxis from critical theory — and more. In accounting for the conditions of its own production, it becomes 'unalienated knowledge' (Stanley, 1990, p. 13). 'In such work, researchers *pry open social mythologies that others are committed to sealing*' (Fine, 1992, p. 221, my emphasis). Contradictory loyalties are exposed. Gender is no longer a unifying concept. Nor is collaborative action research. Nor is abstract knowledge. Nor is the 'reward' of breaking the silences for all participants. Magda Lewis (1992) reminds us of one other caution for this work, a caution which is hard to find in non-feminist critical theory.

> We cannot expect that students will readily appropriate a political stance that is truly counter-hegemonic, unless we also acknowledge the ways in which our feminist practice/politics creates, rather than ameliorates, a feeling of threat: the threat of having to struggle within unequal power relations; the threat of psychological/social/sexual, as well as economic and political marginality; the threat of retributive violence — threats lived in concrete embodied ways. (p. 187)

Critical legal scholar Regina Austin (1989) problematizes the cost of breaking the legal silence surrounding African-American women's bodies within white men's law. Though it is hard for many of us to accept the risks which come from adopting such a position — e.g., the professional vulnerability it occasions, particularly since such positions are easily dismissed by less marginalized academics — some of us might resonate to her words:

> When was the last time someone told you that your way of approaching problems . . . was all wrong? You are too angry, too emotional, too subjective, too pessimistic, too political, too anecdotal and too instinctive? How can I legitimate my way of thinking? I know that I am not just flying off the handle, seeing imaginary insults and problems where

there are none. I am not a witch solely by nature, but by circumstance, choice, [and gendered label] as well. I suspect that what my critics really want to say is that I am being too self consciously black (brown, yellow, red) and/or female to suit their tastes and should 'lighten up' because I am making them feel very uncomfortable, and that is not nice. And I want them to think that I am nice, don't I? (Fine, 1992, pp. 221–2)

Michelle Fine concludes her analysis of using activist research to reframe feminist psychology by asking us to take more risks in our research. To state upfront and throughout our research projects that we are speaking with 'partial knowledges.' To take acknowledged, contradictory, argumentative, and shifting positions in research 'positioned explicitly with questions and not answers; as mobile and multiple, not static and singular; within spaces of rich surprise, not predetermined "forced choices;" surrounded by critical conversation, never alone' (Fine, 1992, p. 230). To do so, we have even to expose our own vulnerabilities — and challenge our supporters.

> The confessional impulse of fronting up with one's position is as important to feminism[s'] commitment[s] to standpoint as it is to masculinist claims to discursive authority across disciplines in the academy . . . the moment women open their mouths . . . they are asked whose name and from what theoretical standpoint they are speaking . . . and to show their identity papers. Our identity papers are dog-eared passports, marked by entry and exit stamps among foreign discourses and languages in which we have traveled as tourists. We speak here at a moment in history, in a language and textual terrain not of our making. We may claim affinity with poststructuralism, but we do not owe it a debt of gratitude. Instead, we salute feminists past and present. After all, it is the voluminous feminist literature of the last two decades that has made the most powerful contribution to re-thinking the subject, to questioning theory in all disciplines, and to the debates on difference. (Luke and Gore, 1992, p. 4)

Such analyses, both painful and full of potential, have deeply colored my collaborative work with teachers engaged in action research. The latest piece of that work was recently published by Teachers College Press (Hollingsworth, 1994c); it is entitled *Teacher Research and Urban Literacy Education: Lessons and Conversations in a Feminist Key*. In the book, I attempt to braid the lessons of feminist research, postmodern discourses, the praxis of action research, and the personal and professional risk needed to take a stand for socio-political changes in urban schooling. The teachers who collaborated with me each took different risks and spoke in contradictory voices throughout in an effort to 'tell it like it is' (or at least like it was when we told it). Let me quote from the preface to the book, so you can see what I mean.

> This book begins with a narrative about a group of teachers and me
> as one of their teacher educators who spent six-plus years together
> learning to teach and conduct research on teaching. It ends with the
> transformation of the teacher educator's research and practice as a
> result of our sustained conversations. We have made an attempt to
> include all of our voices in this text with respect to our various stand-
> points and world views . . . For example, we found that we were
> deeply connected by our passion and commitment to public urban
> education in schools characterized by ethnic diversity, limited eco-
> nomic resources and locations in high crime areas . . . [However] as
> we reflected on our common educational interest, we also realized
> that we had come to that commonality in various ways. (Hollings-
> worth, 1994c)

For example, while (I think) most of us converge at least in the spirit of
feminist perspectives to our work which I will explicate in the text, two points
are important to note: (1) we have jointly constructed these understandings
through our experiences, research and conversations; and (2) we come from
different standpoints on the degree of congruence with feminist epistemology
in the way that I, as the story narrator, have explicated it in the text. Lisa,
Leslie, Jennifer, and Mary have come to basically agree with my position.[3]
Karen is less comfortable with the term 'feminist.' She does not want our
choice of language to alienate those in power who we might influence with
this work. While we understand the political importance and caring stance of
Karen's position, some of us don't worry about upsetting others so much as
we want to provide a language that will reach other women and teachers who
have previously been alienated from more socially polite texts. I also want to
retain the spirit of our varying feminist views to both validate other teachers'
experiences and celebrate the work of those pioneering women and men who
want all experiences and occupations to be valued in this culture, and who are
brave enough to risk personal rejection to do so. Anthony identifies himself
easily as a feminist and is conscious of the role he plays as a white male
teacher in a predominantly female profession. He often reminds us that feminist positions readily transcend issues of gender.

> Inside and outside of Karen's cabin, we [also] dealt head-on with
> issues of race, class and culture. We spent a great deal of time talking
> about how to characterize the differences between ourselves (as vari-
> ously privileged individuals) and the urban public school environ-
> ments in which we work and study and learn to teach. As you'll read
> in [the book], everyone except me vehemently rejected the descriptor
> of 'lower class' to describe our school communities. I wanted to speak
> clearly about the class differences that are usually silenced in discus-
> sions of teacher education. The others felt the term was non-feminist,
> hierarchical, and discriminatory. As we talked, we came to appreciate

each other's points of view, and even shifted a little, though we never found a common language to express our ideas. . . . In short, the discussion about class, race, privilege and luck was an uncomfortable one. All the others in our group disagreed with Jennifer and me. Mary, whose position in a year-round school required her to teach the day following this discussion, felt the need to withdraw from the intense conversation and return to something concrete. Finding our extended discussion both depressing and distracting from her business at hand, she folded laundry and made lesson plans, while the rest of us continued to talk about our varied understandings of the communities from which we came and in which we taught. Though readers of this narrative will not find a restful order or consensus in our words, what is present is an attempt to include all of our perspectives, with all of their complexities and ambiguities. (Hollingsworth, 1994c, pp. ix–xii)

Merging Feminisms and Action Research

So, I've described how feminist approaches to educational research are varied, situated, contradictory, self-critical. I advocate, in my own work, a merger of feminist theories and action research because this seems to push action research deeper into the underlying issues which hinder reformational school reform (such as agencies and epistemologies) and keeps us from a too-comfortable reliance on generalized educational theories. (See Soltis, 1994, for the difference between reform and reformation.) And, the self-contradictory nature of feminist theory itself helps to keep me grounded. Yet, just as feminist interpretations of societies differ within the United States and across the world, they appear to share a commitment to social justice and see education as an important site to reconsider the exclusion of women and other oppressed groups (Weiler, 1993). Because the merger provides both theoretical ground and a grounded project to critique the very 'knowledge' of the purposes and results of education, it also gives me and my students stronger senses of agency and direction.

To construct a case for the power of the merger outside of my own work, I will draw upon two current historical trends which seem to relate to the feminist theories reported in the literature above: current (conservative) constructions of schooling and postmodern views of knowledge. I will close by speaking to the potential revolution in teaching and teacher education which could occur from the intersection of feminist theory and action research.

The Conservative Nature of Schooling

Hugh Sockett and I argued, in the 93rd Yearbook of the National Society for the Study of Education, entitled *Teacher Research and Educational Reform*

(1994), that the critical feminist stance which grew out of the US Civil Rights and Women's Movements have contributed to a contemporary understanding of teacher research in the United States as both a grass-roots and academic revolt against a neo-Conservative attempt to control education. The text is part of a fuller explication of the history of action research, which I have not addressed in the current paper.

> Beyond the notions of teacher research for curricular improvement and reflective practice for pedagogical improvement, a third derivative of action research — critical praxis — is important background to a contemporary understanding of the movement in the United States . . . Reflexively fueled by the Civil Rights and Women's Movements during the 1960s and 1970s, philosophers, scientists and teachers questioned the firm, modernist belief in rationalist science — and the unfragmented and politically silencing nature of knowledge that 'science' assumes. Even popular teacher-promoted curricular projects which challenged static views of knowledge and societal norms — but from a singular philosophic stance, such as the Bay Area Writing Project — were not free from scrutiny (see Delpit, 1986). (Hollingsworth and Sockett, 1994, pp. 7–8)

The social-emancipatory movements growing out of the 1960s and 1970s, however, were certainly not the first to critique the conservative nature of schools. Critical projects against models of schooling along factory and business lines in the United States have been questioned since their inception. Margaret Haley was an elementary school teacher and organizer for the women-populated Chicago Teachers Federation, an organization which protested against not only wage and working conditions, but raised questions of what constitutes an education in a democracy. Others, such as Ella Flagg Young, first woman superintendent of schools in Chicago, and university experts such as John Dewey, challenged educational efficiency experts 'whose greatest concern was cost efficiency and the provision of a standard minimum education for the poor' (Weiler, 1993, p. 212). World War II effectively ended those progressive educational protests.

In the 1980s, it was the argument of organized women and minorities around assertions of non-white, non-western and non-male values and knowledges in the United States which strongly influenced a backlash of right-wing politics. The conservative agenda for education can be seen in the education 'crisis' of the 1980s from a market perspective.

> The reforms of education. . . are again framed in terms of a model of competition and efficiently taken from the corporate world . . . An example . . . can be seen in the 1991 Bush administration's version of conservative educational reform, called America 2000. This plan entailed three main features for public education: (i) parental choice . . . in

which children are envisioned not so much as little workers as little consumers, who would compete for the best educational product . . . Not only would the most disadvantaged children be left in the most disadvantaged schools in this conception, but the ideal of a common, democratic education in which children of different backgrounds would go to school together, is completely absent . . . ; (ii) the promise of $150 million from private corporate sources to design 535 innovative new schools one for each congressional district; and (iii) the introduction of national standardized tests for all children at fourth, eighth, and twelfth grades . . . For all the talk of choice, innovation, and freedom as goals, the real vision of education that underlay these proposals was a narrow vision of fragmented knowledge that could be measured by multiple choice tests . . . Moreover, the tests, to be used to compare students and schools nationwide, would not be accompanied by any greater resouces to provide equal educational opportunity for all students. So once again the advantaged would be competing with the disadvantaged, with almost certainly predictable results. (Weiler, 1993, pp. 218–19)

You might notice that this phenomenon has not been limited to the United States. While progressive theorists and activists have been increasingly on the defensive in the 1980s in the US, Britain, Canada and New Zealand, with stronger left and progressive political traditions, have not been spared from neo-Conservative attacks on the ideals of equality and justice. 'The impact of Thatcherism in Britain described by Madeline Arnot and Gaby Weiner, for example, is different yet familiar to those of us in Canada, New Zealand and the United States' (Weiler, 1993, p. 216; see also Arnot, 1993, and Weiner, 1993). Gaby Weiner identifies a 'regressive modernism' in the Thatcher administration, similar to America 2000, as it faces two directions simultaneously.

Modernism [is an attempt] to replace the 'outdated' approach to social welfare of the post World War II period with free market policies perceived as more in line with the needs of the modern state. Yet at the same time there is a harking back to a golden Victorian Age where subjects were clearly defined and social science was an untried newcomer. (Weiner, 1993, p. 90)

Although not openly 'anti-feminist' (in comparison with the Reagan administration in the US), Mrs Thatcher's Government encouraged the view that it would (similarly) restore patriarchal and 'family' values.

Certainly, the ongoing projects of critical action research projects are helping to bring such matters to light in both the United Kingdom and the United States. Yet there is a failure, particularly in Britain, to *note the interconnections between the rise of the radical right and critical feminist traditions in education* (Arnot, 1993, p. 186).

According to Jane Kenway, mainstream policy analysts in Britain

> ... have rather arrogantly failed to notice that they (most often men) write largely for and about men. Insensitive to matters of gender, they have little or no apparent consciousness of how gender inflected are their theories, concerns and interests. [Further,] many mainstream/ malestream policy analysis seem unaware of an increasing body of feminist scholarship which both exposes many of the limitations of the presuppositions of the policy field and brings matters of gender into the foreground. (Kenway, 1990, p. 7)

Such analyses in Great Britain are remarkably similar to the 'gender-neutral' market theories in the United States. In both countries, an articulated politics of difference, commitments to the uncomfortable and shifting voices of those engaged in the work of society, and a recognition of the relations and positions of power in society which are often part of feminist approaches to educational discourse, pedagogy, and research, would certainly amplify the work of action researchers.

Postmodern Views of Knowledge

A second argument for integrating feminist theories into the work of action researchers has to do with current constructions of 'knowledge.' Since most educators are probably familiar with this phenomenon, I will not belabor the point here. I will simply restate Dorothy Smith's (1987) point that knowledge, *the* economic commodity produced by academics, is constitutive of relations of *ruling* as well as of relations of *knowing*. Though other critical sociologists and political scientists have spoken to us about the problematic nature of knowledge as a political act (see especially Bourdieu, 1977; Foucault, 1980), those challenges did not resonate with my own experience until I heard them articulated from feminist stances. The omission of the gendered experience left me reading (and feeling) Bourdieu and Foucault and others as though I (as woman) still did not have a right to know, to speak, to question, to challenge, to change. The generic 'man' in mankind, and humanist, and humanity, had not historically included me (as child-abuse survivor, female student who wanted to study mathematics, wife and mother, single female parent, woman professor, lesbian community member). Like many black American women's critiques of feminism, I was not certain, without specific naming of my own experience, that I would be included in the 'new' conception of knowledge. As long as my particular woman's viewpoint was seen as too threatening to bring into the conversations *on knowledge*, rather than illuminating the potential for *reconstructing knowledges*, I was not sure at all that my knowledge would count. That my teaching and my research would count. Further, I was not certain that children and students, unnamed in most poststructuralist critiques of knowledge, would be any better off than they were before. Finally, I have noticed throughout my career as a teacher educator that *unless teachers examine and name their own knowledges and questions, they have difficulty becoming critical consumers of other-formulated knowledge.*

Throughout my history, most of my personal experiences in education were based on policies which suggest a 'more or less untroubled reliance on the paradigms of mainstream science and the benign consequences that should follow from their use' (Greene, 1994, p. 423). This has been the case, even recently, in spite of the growing 'acknowledgment of the importance of perspective and point of view in educational inquiry: Gender, class and ethnic identity have been found to play important roles in identifying and accomplishing such purposes as those named by the contributors to *Improving Education*' (Greene, 1994, p. 424).

Most postmodernists in the United States critique the metanarratives written by others which have controlled our collective education — such as those of Bloom (1964), Tyler (1949), Schwab (1972), Shulman (1987), and even Schön (1986). (See Cherryholmes, 1988, for a partial overview.) Poststructuralists are sometimes self-critical in critiquing poststructural models of knowledge against more structured approaches (again, see Cherryholmes, 1988). And yet, they rarely name their own personal positionings inside of their theories: their struggles with pedagogy and ideology, how their students and peers have turned their theories upside down, their own doubts about the validity of their positions. Nor do they ask me about similar occasions and positions in my lived theories. Thus, those seamless depictions of poststructural theory leave me skeptical and fearful. Critical theories could also leave me feeling as if I hadn't found the 'right' theory yet, except for the continual messages I've received across my life that there is no 'right' theory. In a very telling moment as a religious young woman of twenty-two, I went to my Anglican priest for help to restore my faith in a unitary creator. 'I'm beginning to doubt my faith in God,' I confessed. The priest responded not as I'd expected, but with a response which validated my uncertainty and shaped the questions of my life: 'You know, I think I am too.' That early example of what I'd now call a feminist postmodern contribution to eroding the 'theorizing subject' left me with a sense of agency. It still does. Similar experiences have happened to others engaged in feminist action research.

The distinctions between poststructural and feminist theories are of special importance, since it took so long for feminist theorizing to be recognized. Who speaks, after all? Where is the point of departure when it comes to knowing? Who speaks for whom? For Judith Butler, the subject is constituted at certain points where power overlaps action and knowing. Never predetermined, these subjectivities identify various theoretical positions, 'working the possibilities of their convergences and trying to take account of the possibilities they systematically exclude' (Butler, 1990, p. 9). And, as subjects whose power shifts as we are repositioned daily in our collaborative work with others, we can continually remember *both how it feels to know and not know.*

*The Potential for Re-visioning Revolution in Both Teacher
Education and Schooling*

The imagined promise of a feminist project in action research is not only that
teachers will reflect upon their work and responsibilities as professional educators, in conjunction with university collaborators or not, but that each will
speak from their many positions in education, naming and renaming the project
as it comes to light, focused on radical social change, strong in the senses of
both agency and self-critique. Thus feminist notions of self-positioning and
self-critique and continuous challenges to (even our own) constructed metanarratives, offer a strong reinterpretation of action research in both university
and school organizations and practices. Without us all naming our shifting subjectivities and the political powers that control our relations, our values, our
market productivities and reproductivities, we cannot overcome the obstacles
to revolution in our field.

How has my discovery of feminist readings of action research affected my
current sense of self — as teacher educator, academic, action researcher and
critical friend? I no longer need to seek authoritative permission to construct
and critique the possible in educational reform.[4] I still need to learn more
about respectful collaborative relations — recognizing when others are not
ready or willing to take the risks I do. I am often regarded as too impatient and
too impassioned. Yet, I believe in the project. I have witnessed radical changes
take place in teachers' self-descriptions, their urban school classrooms, schools,
and districts. I've seen teachers devise new pedagogies for bringing their children's lives into the classroom and curriculum. I have seen school boards and
state policies change because of feminist action research.

And I am (today) in the company of good and critical friends. I find new
friends who hear me, and then invite me into their 'spaces' to continue the
conversation. The price for this work, though, is certainly estrangement and
strange reconnections. The teachers with whom I work have already outgrown
me in their praxis. Karen Teel won a Spencer Fellowship for teacher research.
Mary Dybdahl is planning an American Educational Research Association session next year with her students and without me. Leslie Minarik is now interested in school restructuring. Anthony Cody is taking a leave from the classroom
to collaborate in a university science project. I'm heading for Asia for two
years. Our work has expanded in different directions — as it should — but I
often miss the dinner discussions with them at my home in Berkeley.

Someone recently asked me, 'If these feminist action research projects
succeed, what will teacher educators do?' I take comfort in the closing of
Michelle Fine and Pat Macpherson's story of their communal and contradictory
explication of feminist experience with a group of teenagers. The authors
learned (and reported) that their guests were not the configuration of their
fantasies for female adolescence, representing the 'after' picture which depicts
all the gains of the women's movement. Their embodied (and heard) stories
shattered the 'hoped for' ideological framing of the research dinners. They

were women of a new generation, challenging our visions of feminism and appropriate action.

> After our last dinner, stuffed and giggly, tired but still wanting just one more round of conversation, we — Pat and Michelle — realized that the four young women were getting ready to drive away. Together and without us . . . We turned to each other, realizing that even our abandonment was metaphoric and political. These four young women were weaving the next generation of feminist politics, which meant, in part, leaving us. We comforted ourselves by recognizing that our conversations had perhaps enabled this work. (Fine and Macpherson, 1992, p. 202)

I invite you to share in the potential of your own contributions and engage with me in conversation about the possibilities for the intersection of feminist theory and action research, and the resultant opportunities for societal, as well as educational change.

Notes

1 Consideration of the social, political and epistemological implications of public school teaching as a profession heavily populated by women (particularly at the elementary level) will be addressed throughout this paper.
2 The historical gap between the first wave of feminism, usually associated with women's right to be educated, and the second wave — associated with the right to critique knowledge and become educational administrators — leaves invisible the feminist work which occurred in between (see Rupp and Taylor, 1990).
3 Mary explained: '*I have identified myself as a feminist for most of my adult life. Through this group, I have come to see the importance of articulating this way of thinking as a valid / important way to approach teaching.*'
4 A colleague of mine, Diane Holt-Reynolds, told me at a recent American Educational Research Association meeting that I acted as though I were 'entitled' to knowledge.

References

APPLE, M.W. (1985) 'Teaching and "women's work": A comparative historical and ideological analysis', *Teachers College Record*, **86** (3), pp. 461–81.

ARNOT, M. (1993) 'A crisis in patriarchy? British feminist educational politics and state regulation of gender', in ARNOT, M. and WEILER, M. (eds) *Feminism and Social Justice in Education*, London, Falmer Press, pp. 186–209.

ARNOT, M. and WEILER, K. (1993) *Feminism and Social Justice in Education*, London, Falmer Press.

AUSTIN, R. (1989) 'Sapphire bound!', *Wisconsin Law Review*, No. 3, pp. 539–78.

BEAUVOIR, SIMONE DE (1949) *The Second Sex*, New York, Bantam.

BELENKY, M.F., CLINCHY, B.M., GOLDBERGER, N.R. and TARULE, J.M. (1986) *Women's Ways of Knowing: The Development of Self, Voice, and Mind*, New York, Basic Books, Inc.

BELLAH, R.N., MADSEN, R., SULLIVAN, W.M., SWIDLER, A. and TIPTON, S.M. (1987) *Individualism and Commitment in American Life: Readings on the Themes of Habits of the Heart*, New York, Harper and Row.

BIKLIN, S.K. and POLLARD, D. (1993) *Gender and Education*, 92nd Yearbook of the National Society for the Study of Education, Chicago, University of Chicago Press.

BLOOM, B.S. (1964) *Stability and Change in Human Characteristics*, New York, Wiley.

BOURDIEU, P. (1977) 'Cultural reproduction and social reproduction', in HALSEY, H. and KARABEL, J. (eds) *Power and Ideology in Education*, New York, Oxford University Press, pp. 487–551.

BRITTAN, A. and MAYNARD, M. (1984) *Sexism, Racism, and Oppression*, Oxford, Basil Blackwell.

BRITZMAN, D.P. (1993) 'Beyond rolling models: Gender and multicultural education', in BIKLIN, S.K. and POLLARD, D. (eds) *Gender and Education*, 92nd Yearbook of the National Society for the Study of Education, Chicago, University of Chicago Press.

BURKE, C.G. (1978) 'Report from Paris: Women's writing and the women's movement', *Signs: Journal of Women in Culture and Society*, **3** (4), pp. 843, 855.

BUTLER, J. (1990) *Gender Trouble: Feminism and the Subversion of Identity*, New York, Routledge.

CARR, W. and KEMMIS, S. (1986) *Becoming Critical: Education, Knowledge and Action Research*, London, Falmer Press.

CHERRYHOLMES, C.H. (1988) *Power and Criticism*, New York, Teachers College Press.

CHODOROW, N. (1978) *The Reproduction of Mothering*, Berkeley, The University of California Press.

CLIFFORD, G.J. (1989) 'Man/woman/teacher: Gender, family and career in American educational history', in DONALD, J.J. (ed.) *American Teachers: Histories of a Profession at Work*, New York, Macmillan, pp. 293–343.

DAVIS, A. (1981) *Women, Race, and Class*, New York, Random House.

DELPIT, L.J. (1986) 'Skills and other dilemmas of a progressive Black educator', *Harvard Educational Review*, **56** (4), pp. 379–85.

ELLIOTT, J. (1991) *Action Research for Educational Change*, Milton Keynes, Open University Press.

ELLSWORTH, E. (1989) 'Why doesn't this feel empowering? Working through the repressive myths of critical pedagogy', *Harvard Educational Review*, **59** (3), pp. 297–324.

FINE, M. (1992) *Disruptive Voices: The Possibilities of Feminist Research*, Ann Arbor, The University of Michigan Press.

FINE, M. and MACPHERSON, P. (1992) 'Over dinner: Feminism and adolescent female bodies', in FINE, M. (ed.) *Disruptive Voices: The Possibilities of Feminist Research*, Ann Arbor, The University of Michigan Press.

FOUCAULT, M. (1980) *Power/Knowledge*, New York, Pantheon.

FREIRE, P. (1970) *Pedagogy of the Oppressed*, New York, Continuum.

GIROUX, H.A. (1988) *Teachers as Intellectuals*, Granby, MA, Bergin and Garvey.

GILLIGAN, C. (1982) *In a Different Voice: Psychological Theory and Women's Development*, Cambridge, MA, Harvard University Press.

GORE, J. (1992) *The Struggle for Pedagogies*, New York, Routledge.

Gramsci, A. (1971) *Selections from the Prison Notebooks of Antonio Gramsci*, Q. HOARE and G. SMITH (eds, Trans.) New York, International Publishers.

GREENE, M. (1978) 'Wide-awakeness and the moral life', in GREENE, M. *Landscapes of Learning*, New York, Teachers College Press, pp. 42–73.

GREENE, M. (1994) 'Epistemology and educational research: The influence of recent approaches to knowledge', in HAMMOND, L.D. (ed.) *Review of Research in Education*, Volume 20, Washington, D.C., American Educational Research Association, pp. 423–64.

HARAWAY, D. (1988) 'Situated knowledges: The science question in feminism and the privilege of partial perspective', *Feminist Studies*, (14), pp. 575–99.

HARDING, S. (1986). *The Science Question in Feminism*, Ithaca, N.Y., Cornell University Press.

HOLLINGSWORTH, S. (1989) 'Prior beliefs and cognitive change in learning to teach', *American Educational Research Journal*, **26** (2), pp. 160–89.

HOLLINGSWORTH, S. (1994a) 'Feminist pedagogy in the research class: An example of teacher research', *Educational Action Research*, **2** (1), pp. 457–79.

HOLLINGSWORTH, S. (1994b) 'Teacher research', *International Encyclopedia of Teacher Education*, Oxford, Pergamon Press.

HOLLINGSWORTH, S. (1994c) *Teacher Research and Urban Literacy Education: Lessons and Conversations in a Feminist Key*, New York, Teachers College Press.

HOLLINGSWORTH, S. and SOCKETT, H. (eds) (1994) *Teacher Research and Teacher Education Reform* (Yearbook of the National Society for the Study of Education), Chicago, University of Chicago Press.

HOOKS, BELL (1986) *Talking Back: Thinking Feminist, Thinking Black*, Boston, South End Press.

HOOKS, BELL (1990) *Yearning: Race, Gender, and Cultural Politics*, Boston, South End Press.

JAGGAR, A. (1983) *Feminist Politics and Human Nature*, Sussex, The Harvester Press.

KENWAY, J. (1990) *Gender and Education Policy: A Call for New Directions*, Victoria, Australia, Deakin University.

KENWAY, J. and MODRA, H. (1992) 'Feminist pedagogy and emancipatory possibilities', in LUKE, C. and GORE, J. (eds) *Feminisms and Critical Pedagogy*, New York, Routledge, pp. 138–66.

KLEIN, S., ORTMAN, P.E., CAMPBELL, P., GREENBERG, S., HOLLINGSWORTH, S., JACOBS, J., KACHUCK, B., MCCLELLAND, A., POLLARD, D., SADKER, D., SADKER, M., SCHMUCK, P., SCOTT, E. and WIGGINS, J. (in press) 'Continuing the journey toward gender equity', *Educational Researcher*.

LABAREE, D.F. (November, 1993) 'The lowly status of teacher education in the United States: The impact of markets and the implications for reform'. Paper prepared for the Rutgers Invitational Symposium on Education (RISE) conference on 'Promise and Challenge in Teacher Education: An International Perspective.'

LAIRD, S. (1988) 'Reforming "Women's true profession": A case for "feminist pedagogy" in teacher education', *Harvard Educational Review*, **58** (4), pp. 449–63.

LATHER, P. (1986) 'Research as praxis', *Harvard Educational Review*, **56** (3), 257–77.

LATHER, P. (1991) *Getting Smart: Feminist Research and Pedagogy with/in the Postmodern*, New York, Routledge.

LECK, G.M. (1987) 'Review Article — Feminist pedagogy, liberation theory, and the traditional schooling paradigm', *Educational Theory*, **37** (3), pp. 343–55.

LEWIS, M. (1992) 'Interrupting patriarchy: Politics, resistance and transformation in the feminist classroom', in LUKE, C. and GORE, J. (eds) *Feminisms and Critical Pedagogy*, New York: Routledge, pp. 167–191.

LORDE, A. (1984) *Sister Outsider*, Trumansburg, NY, Crossing Press.

LUKE, C. and GORE, J. (1992) *Feminisms and Critical Pedagogy*, New York, Routledge.

MARTIN, J.R. (1981) 'The ideal of the educated person', *Educational Theory*, **31** (2).

McKERNAN, J. (1991) *Curriculum Action Research: A Handbook of Methods and Resources for the Reflective Practitioner*, New York, St. Martin's Press.

NODDINGS, N. (1986) 'Fidelity in teaching teacher education, and research for teaching', *Harvard Educational Review*, **56** (4), pp. 496–509.

NODDINGS, N. (1992) *The Challenge to Care in Schools*, New York, Teachers College Press.

ORNER, M. (1992) 'Interrupting the calls for student voice in "liberatory" education: A feminist poststructuralist perspective', in CARMEN LUKE, C. and GORE, J. (eds) *Feminisms and Critical Pedagogy*, New York, Routledge, pp. 167–89.

RAYMOND, J. (1986) *A Passion for Friends: Toward a Philosophy of Female Affection*, Boston, Beacon Press.

REINHARZ, S. (1992) *Feminist Methods in Social Research*, New York, Oxford University Press.

ROBINSON, L. (1989) 'What culture should mean', *The Nation*, September, pp. 319–21.

RUPP, L. and TAYLOR, V. (1990) *Surviving in the Doldrums: The American Women's Rights Movements, 1945–1960s*, Columbus, OH, Ohio State University Press.

SCHÖN, D.A. (1986) *The Reflective Practitioner*, New York, Basic Books.

SCHWAB, J.J. (1972) 'The practical: A language for curriculum', in PURPEL, D.E. and BELANDER, M. (eds) *Curriculum and the Cultural Revolution*, Berkeley, CA, McCutchan, pp. 72–79.

SHULMAN, L.S. (1987) 'Knowledge and teaching: Foundations of the new reform', *Harvard Educational Review*, **57** (1), pp. 1–22.

SOLTIS, J. (1994) 'The new teacher', in HOLLINGSWORTH, S. and SOCKETT, H. (eds) *Teacher Research and Teacher Education Reform* (Yearbook of the National Society for the Study of Education), Chicago, University of Chicago Press, pp. 245–260.

SOMEKH, B. (Spring, 1993) 'Thinking about CARN in March, 1993', *CARN Newsletter*, Issue 1, Bournemouth, Bourne Press.

SMITH, D. (1987) *The Everyday World as Problematic: A Feminist Sociology*, Boston, Northeastern University Press.

STANLEY, L. (1990) *Feminist Praxis: Research, Theory and Epistemology in Feminist Sociology*, New York, Routledge.

STENHOUSE, L. (1975) *An Introduction to Curriculum Research and Development*, London, Heinemann.

TYLER, R.W. (1949) *Basic Principles of Curriculum and Instruction*, Chicago, University of Chicago Press.

WALKERDINE, V. (1986) 'Progressive pedagogy and political struggle', *Screen*, **27** (5), pp. 54–60.

WARREN, D. (1989) *American Teachers: Histories of a Profession at Work*, New York, Macmillan.

WEILER, K. (1988) *Women Teaching for Change: Gender, Class and Power*, South Hadley, MA, Bergin and Garvey.

WEILER, K. (1993) 'Feminism and the struggle for a democratic education: A view from the United States', in ARNOT, M. and WEILER, K. (eds) *Feminism and Social Justice in Education*, London, Falmer Press, pp. 210–226.

WEINER, G. (1993) 'Shell-shock or sisterhood: English school history and feminist practice', in ARNOT, M. and WEILER, K. (eds) *Feminism and Social Justice in Education*, London, Falmer Press, pp. 17–100.

WELLESLEY COLLEGE CENTER FOR RESEARCH ON WOMEN (1992) *How Schools Shortchange Girls*, Annapolis Junction, MD, American Association of University Women.

3 Context, Critique and Change: Doing Action Research in South Africa

Melanie Walker

ABSTRACT *While this article is located in a specific country, South Africa, the arguments developed are more generally useful for action researchers internationally. The author first discusses the critical importance of the broader context within which any action research project is embedded. Here, the point is to understand the specificity of the local political, research and educational policy climate, and how one's own biography as a researcher is shaped by this set. Fragments from the author's life, together with three South African action research projects, are then outlined to illustrate these points. The case is then developed for the notion of critique in action research, arguing in effect for the central practical importance of rigorous theory-based projects if action researchers are going to be able to step outside their taken-for-granted reality and push at the edges of their own experiences. The case is supported with reflections on the author's own educational development, shaped over time and grounded in the dialectical play of sociological theories and her empirical texts. Finally, the author considers what all this means for change: for her personal understanding, for wider questions of collective political action and for action research projects.*

Introduction

My understanding of action research in South African settings has developed over the last ten years: first reading about it; then action researching my own role in supporting curriculum development in african[1] township schools; and now in working with lecturers at the University of the Western Cape (UWC), a historically black university[2] struggling to embed a research ethos and culture, and to throw off its apartheid origins as a 'bush' college. My own current interest is in action researching gender, 'race' and educational management and leadership in the academy, refracted through my new location in an administrative role.

Autobiographical Notes

This was how I had intended to start my address to the CARN conference (and indeed did). But informal conversations with Susan Noflke at the American Educational Research Association a few days previously, and listening to Sandra Hollingsworth's address at CARN on feminist action research, troubled me into puzzling over my own gendered and raced positionality. Now I must try to capture in this brief text what I added after my introductory paragraph on a wonderfully sunny if cool spring day at the University of Birmingham, on the eve of the democratic elections in South Africa to which I was to return immediately after the conference in order to cast my own vote. This in itself, coupled with extensive coverage in the British press and on television in the few days I had already spent in England, had further intensified my own emotional responses and feelings of furious hope which were not to be dampened even by the bomb blast soon after my arrival at Jan Smuts airport in Johannesburg.

What then did I try to convey to the CARN audience that would position me more firmly in relation to my text? I began by mentioning the question that had troubled me while listening to Sam. Why was gender only now becoming an explicit research and teaching issue for me? Was I like the woman she had quoted in her paper who, try though she would, simply could not feel oppressed? Why had critical theorists like Freire and Giroux resonated for me when I had encountered them in the 1980s? Why had I not felt excluded from their texts as Sam had?

To answer these questions I had to tell a fragment of my own life, growing up white in a South Africa dominated by National Party rule and racial laws, in Natal province which saw itself as the last outpost of the British Empire, and the city of Durban where adult African men and women were named 'boys' and 'girls', and where buses, park benches, entrances to post offices and residential areas — everything was segregated. In my family, politics was seldom discussed and questioning of my father's authority was firmly discouraged. I recall muted rumblings of fear around the time of the Cato Manor 'riots' (a black township near Durban) in 1959, the causes and history of which I was only to understand many, many years later. I went to segregated schools and a single sex girls' high school (a British ideal). Growing up in a middle-class home, insulated in a comfortable suburb from the hazardous black urban locations, unaware and unconcerned about the desperately unfair irregularities of South African society, the only black people I encountered were the two domestic workers in my home, intimate strangers.

I started university in 1969 in Durban. By 1970 the handful of Indian students had left the campus, required by law to attend the new University of Durban-Westville intended by apartheid planning in higher education for them. But it was also in 1970 that I was finally to learn another side of South African history in a Political Science course taught by a gifted and generous researcher, Rick Turner (later to be assassinated by mysterious and still unknown assailants).

I found out about the injustices of the Land Act, I participated in my first protest march, I understood my first organizational work as we attempted to enlighten Durban's citizens about the cruelty of the detention laws.

Yet I cannot claim that from this point I was active in working politically for change (many others were). I railed privately against the injustices of South African society and argued vehemently with family members. But it was only some years later, influenced after a stay in London by Marxist interpretations of the past and then taking up a post as a history teacher in a 'Coloured' school near Cape Town, that I began to act in my classroom for change. My political education was accelerated by the schools' boycotts of 1980 in which my students rejected 'gutter education' and 'white education' as both contributing to the maintenance of the capitalist order. It was also around this time that I became active in youth and civic politics and, from 1985, in a teachers' union. All these experiences informed my reading of texts which I searched out in order to understand better the situation in which I found myself. I guess I was not looking for myself in those texts, but for a theoretical analysis that helped explain the race and class oppression of my students. While, like Sam, I longed for the critical education I never had, I had been vastly more fortunate simply by virtue of my white skin and the quality of education automatically available to me. Complaining about my own gendered oppression as white and middle class simply wasn't that important, although at times I asserted myself in working relations with male colleagues in ways which earned me the reputation of being 'difficult' and 'aloof'.

Now it seems to me that the emergence of this new South Africa is my liberation too, allowing me to pursue a feminist trajectory which places gender (without deleting race and class) at the heart of my own agenda and new narratives of action research. I do not believe that action research can liberate participants in a 'grand' sense. The real responsibility is to change oneself, searching and struggling with others for the social spaces in which we might challenge and reassemble the self, to see more intensely, to turn easy answers into critical questions, and to practice freedom by recognizing recurring games of truth through a critique always of 'what is' (Foucault, 1980, 1981). This includes problematizing my own university and its rules and practices — an aspect often strangely absent in action research by academics, whose concern for context seems to be more often directed outwards, away from the transformation of their own institutions.

My intention is to set action research in the context of South Africa in volatile transition and the prevailing educational and research climate. My curiosity will be to try and chart some of my own theoretical shifts, shaped by autobiographical factors and the kinds of intellectual influences to which I have been exposed, living as I do 'on the margins' in a global world where intellectual production and research publication is dominated by the North. These experiences will inform my exploration of the 'potential theoretic competence of social actors' (Winter, 1987, p. 5), which I understand as the relationship of self–other and the ways in which we might shift from commonness to critique, from self-reproduction to self-transcendence. In all this I take seriously John Elliott's (1991a) injunction that anything that helps us to make wiser and better judgments about practice is important.

Context

South Africa in Transition

Without doubt, it is not easy at times to be sanguine about South Africa's future, notwithstanding the goodwill and optimism generated by the first democratic elections. Despite the recent political shifts, it is still the most unequal society in the world among countries for which national household income is available.[3] In 1991 the poorest 40 per cent of households in South Africa earned a mere four per cent of the total income; the richest 10 per cent earned more than half (*Weekly Mail,* 11–17 March 1994). All this is concretized in bleak and grim lives: massive adult and youth unemployment; burgeoning urban squatter settlements without access to clean water or adequate sanitation; staggering crime statistics; with one person dying from Tuberculosis every forty minutes.

Yet this is also a society moving away from a ravaged and deeply scarred apartheid past and from an education system which deliberately sought to stifle intellectual capacity and one which is moving toward policies based on principles of democratic participation, a redress of imbalances of access and power, and an overall reconstruction of the education system to develop human potential (National Congress, *A Policy Framework for Education and Training,* January 1994). This is a far cry from the dark days of Verwoerdian[4] grand apartheid and Bantu education to teach africans their 'proper place':

> If the native in South Africa today . . . is being taught to expect that he [sic] will live his adult life under a policy of equal rights, he is making a big mistake . . . There is no place for him in the European community above the level of certain forms of labour. (Quoted in Rose and Turner, 1975, p. 99)

Crudely stated, there was then 'no use in teaching a Bantu child mathematics when it cannot use it in practice'. It then followed, said Verwoerd that: 'People who believe in equality are not desirable teachers for natives' (quoted in Harsch, 1980, p. 99).

Research Context

These remainders of the recent past stand in sharp contrast to African National Congress (ANC) policies in which teachers are conceptualized as 'competent, confident, critical and reflective'. In this new context, then, action research might contribute to working out locally a practical vision of different forms of educational practices as we build a democratic South Africa. But it must necessarily also be contextualized against the prevailing research climate in South Africa, one which is not particularly conducive to educational research in general, even less so to action research. South Africa has one of the lowest ratios of researchers per head of population in the world. Our educational research traditions are fragile, distorted by a past in which Conservative, white Afrikaner intellectuals supported the social engineering of apartheid by constructing educational philosophies which justified segregation and the domination of white over black, all dressed up in the pseudo-scientific language of 'fundamental pedagogic'.[5]

Yet progressive educational research has not served practice particularly well either, given that the dominant discourse here has been located at the macro level, turning above all on the relationship between educational change and social transformation, and more recently on macro policy formulation. Influenced by the key political strategy articulated in the South African Communist Party's 1976 document, 'South Africa — No Middle Road', any reform of capitalism was regarded as impossible, leading to a polarization prior to 1990 of reform and revolution, as much in education struggles as political organization. This resulted in a simplistic drawing of linear conclusions in this macro discourse so that a concern with everyday realities of teaching and classroom processes was devalued as merely 'reformist' in its apparent failure to articulate directly with mass organizations.

Despite the repeated emphasis on the deplorable conditions of teaching and learning, the effect, then, has been a neglect of schools and classrooms as a serious locus for research and a dismissive attitude to the potential of action research in the past. Indeed, action research was criticized by a leading South African scholar of education and left wing intellectual-activist for its narrower concern with transformations in the classroom, thereby 'turning inwards to the educational process as such, divorced from its location in the broader, dominant social structure' (Wolpe, 1991, p. 81). Others have argued that quantitative and qualitative research 'of any sophistication' in and on schools is virtually non-existent, given problems of access, race, gender, class and university of origin (historically black or historically white) of researchers (Chisholm, 1992, p. 158). Add to this that action research has been contemptuously described in the leading local educational journal as a 'facet of "politically correct" instruction', 'mere technique' and 'theoretically . . . very small beer' (Appel, 1991, p. 104).

Certainly, the prospects for action research taking root appear daunting where teachers under conditions of apartheid education simply have not seen themselves as agents in curriculum development or educational knowledge producers. The point here is not to underestimate the damaging effects of over four decades of an education system designed to deliberately stifle intellectual development. Thus, teachers in South Africa mostly do not start from positions of innovative and reflective practice, given the history and effects of author-itarian surveillance of teachers' working lives, of political oppression, and a truncated view of their professionalism which has turned on teachers as mere instalments of state ideology.

An Action Research 'Community'

On a more positive note, however: an action research network has recently been established with its locus at the University of the Western Cape (UWC) (see Davidoff, Julie, Meerkotter and Robinson, 1993); an action research Mas-ters programme at UWC with a clear emancipatory intent has been in exist-ence since 1987; postgraduate courses at the Universities of Cape Town and the Witwatersrand are contributing to developing action research through award-bearing courses; a few non-government organised projects have explored ac-tion research; and a handful of doctoral studies by teachers, teacher educators and health practitioners further contribute to action research theory and practice. And of course, it is precisely this range of action research studies which ex-plore processes of educational change in real settings in ways which provide policy makers with a window onto the conditions under which teachers teach, the same conditions under which students are expected to learn. These studies incorporate marginalized subjects and voices and ground our understanding of educational change in research-based knowledge. Just three illustrative exam-ples will suffice, taken from work in african primary schools, in a non-racial private secondary school and a postgraduate course for teachers at UWC.

Three Action Research Studies

The first example is situated in teacher development work in mathematics classrooms in some of Cape Town's poorest schools. South Africa produces a pitifully small number (some 500) of pupils from african schools who pass matriculation mathematics, but there is also an increasing emphasis now on the need for science and technology policies that promote economic develop-ment. What might implementation of such policies mean for maths lessons in real classrooms? How much do we know and understand of what such shifts demand? Wendy Colyn (1992) researching her practice as a facilitator explains that her insights 'were not gained in the libraries, nor were they gained in the seminar rooms but through work in classrooms and schools' (p. 90). To this end, she describes a lesson with a class of fifty-nine Year 5 pupils,

> [The teacher] feels a sense of urgency. A 30 minute period and so much to do. She starts her explanation in English. She then translates it into Xhosa as she is not certain that the learners have followed in English. As usual more than half the Maths lesson becomes an English

lesson. The learners are asked to repeat nearly every phrase the teacher uses. Their replies are voiced rhythmically and automatically. She asks a few questions, discovering that only a few children have even followed half of what she has been explaining . . . Mrs M glances at her watch. Time is running out so she sets the learners three problems based on the lesson. While they are busy with problems, she goes around to desks and starts to sign the work they are doing. There is no hope of thoroughly marking 59 books. Even if she had time to notice that Nosipho now in year 5 at the age of 14 still has no idea what the difference is between multiplication and division — so 4×2 is the same to her as 4 divided by 2 — what could she realistically do about it? There are 59 learners in the class and many of them do not know the very basics of Mathematics. Bongiwe, the top maths achiever in the class, has completed her work and starts to trace out the letters of her name on the back cover of her book. She yawns as she idly colours the letters. Sindisuxa, thankful that Mrs M is starting at the other end of the class, takes out her Xhosa grammar book — now is the time to catch up on the homework she did not do. Maybe she can avoid Mrs J and the piece of PVC piping that can burn and hurt your hands . . . Behind her Monde is only now finishing the date and preparing to start the exercise. (1992, pp. 96–7)

Having such pictures of classrooms sketched for us is not to pathologise teachers working under dreadfully difficult conditions, nor to lose sight of a democratic educational vision of access for all children to maths and science. Rather, it cautions us to pay far more careful attention to the implementation of new policies in ways which include these pupils, this teacher and this school.

The second example is drawn from Brenda Leibowitz's (1991) classroom-based study, focusing on the learning experiences of her Year 11 English class. Her study emphasizes the need for research on multilingual education in South Africa to document and assess schools' responses so that policies can be formulated which benefit pupils. She describes her own position:

From a practical point of view, when the year began I was almost panic stricken at the prospect of having to face a class containing many different academic levels and in which the L2 [second language] students were evidently too intimidated to speak out or ask questions. I therefore had to devise a way of encouraging the participation of all students in the class, and of catering to the academic and linguistic needs of all the students. (p. 18)

Students also commented on their prior experiences of learning English:

Some of us, English is the second or third language. This language is not easy even though it looks that way. We struggle to learn it but people fail to understand the way we struggle. We are then told *we are slow* thinkers. That puts a person off. (Quoted in Leibowitz, 1991, p. 21)

They also spoke out on what it meant to them to participate in a multilingual, non-racial class where stereotypical views of others break down:

I think it's nice being in a mixed class. I mean at a coloured school, you sort of tend to think that all white people tend to think the same way, or all african people think the same way and it isn't so. I mean you learn a lot of things about them . . . when I'm in the class I don't think of Bruce as a white boy, or whatever, he's just a person. (Quoted in Leibowitz, 1991, p. 20)

There are clear lessons here for a multilingual country striving to put the apartheid past behind us.

The third example is drawn from a study by Mike Adendorff (1994) of the development of a module in 'Radical Pedagogy' for postgraduate teachers. He explores in fine detail the tensions and real constraints between educational practice and social transformation as the module unfolds against the backdrop of political upheaval in late 1989. This, in turn, throws up fascinating questions around voice — who is allowed to speak when there are political differences, not least between the lecturer and his assumptions, and the position of some students who, while committed to resistance politics, were not aligned to the dominant tendency (broadly, the ANC).

This study enables us to access these different views and voices in ways which seem to me to have important lessons for educational processes in a tolerant and democratic society. These students assert their diverse positions in the class in ways which suggest the ambiguities and tensions of this class as a single ideal 'community'. Thus, at one point one of the students comments in an interview:

Why I thought that I had to push the way I see things was that I knew there were a couple of people in the class who were not quite satis-fied with the way things were going . . . what happened here was a preference for the philosophy of the MDM [mass democratic move-ment]. Personally I feel that we should not align ourselves in this situation with a particular political tendency because the Western Cape, much more than the rest of the country, is a heterogeneous entity, where political tendencies are concerned. (Quoted in Adendorff, 1994, p. 256)

What emerged then were multiple communities, complexity and ambiguity, and the web of connections made between competing views:

> What was good for me was that through the course I met people like Stephen L, and we actually, despite our differences, got on. For me, it's quite something to differ strongly with somebody but yet on a personal level to get on. But this doesn't mean compromising your position. (Quoted in Adendorff, 1994, pp. 268–9)

Moreover, given its 'radical' purpose at a time of heightened political activity, it was inevitable that the tensions experienced in the community at large were brought into the classroom, encouraging more thorough thinking through of diverse positions. But in the end, practice also suggested that there were limits to how much 'relevance' and how much integration between a module and 'struggle' was possible — a university course could not of itself generate collective political action.

Critique

> There's a time when people's experience runs out ... (Horton and Freire, 1990)

Important though such studies are, not least in their common concern for social justice, they nonetheless leave a number of unanswered questions relating to our own educational development through action research. I wish to signal and explore just a few of these issues and questions, highlighted in Wendy Colyn's statements which emphasize her experiential learning: is experience enough as the sole arbiter of knowledge? How, for example, do we avoid simply recycling common sense? How do we create new interpretative categories which we might otherwise not encounter? Which theories incite, disrupt and question the taken for granted and even the problematic, and how does the curriculum practitioner access these often esoteric discourses?

My questions, which turn on the notion of the critical potential of theories to help us interrogate practice in a systematic way, seem especially compelling for those wanting to do action research in the new South Africa. The point needing emphasis here is that new education policies, however welcome, have no fixed meaning outside of their critical contestation under the conditions of a negotiated rather than a revolutionary transfer of political power. To this end, we might usefully note the shifting meaning of the term 'democracy' in the current political climate: from the will of the people for the common good, to a new minimalist definition stressing democratic methods and procedures and entrenching free enterprise. In other words, the 'pursuit of the common good' is strikingly absent (Kallaway and Tsibani, 1993).

The challenge then is certainly to ground action research in a discourse of reconstruction, but one nonetheless underpinned by critique, contradiction and contestation over educational and political concerns for social justice. Divorced from such concerns and 'passionate scholarship', action research may easily be domesticated, congealing into a hegemonic orthodoxy which may not best serve the interest of all South African children. Moreover, an engagement with theory seems to me not to be necessarily opposed to a view of our self-understandings of practice generating critical self and social reflection (Elliott, 1991b). I regard such understanding as certainly shaped by practical action, but equally by theoretical encounters, both textual and social, as part of action research communities. Theory, in my view, is not only what is written down.

Still, it may well be that these concerns are generated by my immersion in an academic culture and my present involvement with staff development at UWC (University of Western Cape), with less relevance perhaps to people in diverse practice settings for whom I cannot honestly claim to speak. On the other hand, my position seems not irrelevant to a British audience when a recent editorial in the *Journal of Education for Teaching* (Vol. 19, No. 2) argued for 'a rigorous theory-based system of teacher education' in a climate of 'almost wholly malign policy on teacher education'.

I therefore want to argue that critical, sociological and other theories provide us with categories and frameworks for thinking that enable us to deconstruct common sense and reconstruct it as 'good sense' (Gramsci, 1971). The point is to shift from immediate problem solving to the complexity of critical educational processes, where the latter may not necessarily solve our immediate practical problems, but are likely to generate new questions as we find out what we do not know, rather than what we know (Horton and Freire, 1990). I raise this issue, moreover, in the context of teaching as a kind of bricolage,[6] including teaching in my own university where most lecturers view teaching as mere technique, and low status technique at that. In this view, theoretically informed explanations are discarded (or not even considered) in favour of collecting and improvising practices in an ad hoc, non-analytical search for everyday explanations and methods which work (Hatton, 1988; Dowling, 1993).

What is at stake here is the difficulty of stepping outside one's own taken-for-granted reality precisely because that reality is taken for granted: 'Where I might see sexism in a classroom', Kelly (1985) argues, 'a group of teachers who have never questioned the patriarchal basis of our society will probably not notice it' (p. 144). Yet the situation would be rather different for feminist teachers who begin with a commitment to challenging gendered inequalities and relate research questions to ideas drawn from feminist theory (Weiner, 1989). Similarly, in my experience, if one brings a democratic political commitment to action research, then this political discourse will shape the research. The critical point underscored here is that reflection on practice is not necessarily critical or radical, nor action research inherently transparent (or empowering or emancipatory) outside of its location in particular discourses (Gore, 1993). Thus, 'experience' depends on the discursive conditions of possibility, not least the interpretative frameworks which mediate that experience.

Very recently I have been reading Deborah Britzman's (1991) critical study of learning to teach, and in her account have found further clarity for what troubles me in the 'myth' that experience 'is telling in and of itself' (p. 7). Myth, she says, simply leads us along pre-existing paths, makes available to us only already known practices, and resists explanations 'about the complications we live' (p. 7). Moreover, because we experience teaching individually, we run the risk of failing to situate our personal knowledge in the context of power relations and authoritative discourses. We ought to turn then to theory to politicize common sense, to 'trouble' and 'dispute' its normalizing tendencies to sustain the world as 'given'. In Britzman's view, then, theory does not stand back from or apart from practice, but engages and intervenes. For example, in my own work as a history teacher, I had what Myles Horton (Horton and Freire, 1990, p. 128) has described as the 'right sensitivity' when I became caught up in the 1980 schools' boycott. But it was my reading at that time of Freire's concept of banking education and conscientization (secretly and subversively because his books were banned at the time) and Bowles and Gintis's explanation of the reproductive functions of education that helped me develop a framework to situate schooling in relation to social transformation. My understanding was later refined through accounts of resistance to the domination of the state in educational institutions by writers like Giroux and Apple, together with Gramscian theories on hegemony and political strategy.

More recently, poststructuralist themes on subjectivity, discourse and power–knowledge relations have resonated powerfully for me in revisiting earlier work and better understanding current action. This is notwithstanding my wariness of explanations which, taken to one extreme, seem to justify intellectual disengagement and quietism by relegating political activism to the scrapheap and substituting textual for political radicalism. Admittedly, I find poststructuralist theorizing dense and difficult, to say the least. For the practitioner to move into these theoretical spaces, is nonetheless a worthwhile challenge, and this move must entail one into what Dowling (1993) describes as the 'esoteric' rather than the 'public' domain, if we are not to strip theoretical frameworks of their potential richness for practice. But this is also not to treat theory in some reverential way. Our empirical research texts — interviews, observations, notes and so on — equally work to surprise and interrogate our theory.

Tracking Shifts in my Own Understanding

I want now to refer briefly to my own development, tracking some key theoretical shifts. When I began work with township teachers in what was to be a three-year project in which I would be researching my own practice as facilitator of teachers' development through curriculum changes, one of the aspects I was concerned to find out more about as this work progressed was how teachers experienced their working lives. How did they speak about the institutions that marginalized and excluded them? At that time I was using a framework of ideology critique to analyse what teachers said, and understood ideology as 'false consciousness' so that the teachers held, in my view, a partial understanding of Bantu education's 'true' intentions. Yet, in revisiting my earlier account, now employing Foucault's (1980) analytical tools of the relations of truth and power, I am bound to reconsider the role of researcher-intellectual as one who knows the essential 'truth' hidden below false consciousness. Instead, different notions of subjectivity and discourse deconstruct the idea of a unitary rational agent having some essence or core, independent of experience. In this view, subjects take up different discourse positions, and at times these positions may be contradictory and partial, but not either true or false. The point is that truth claims are historically and socially specific so that the question becomes not what counts as true but why this counts as true. Teachers were indeed conscious of the conditions under which they worked, the circumstances of their lives, the surrounding poverty, and the inconsistencies of the dominant discourse (caring on the one hand, repressive surveillance on the other). What I was doing in ascribing their partial accounts to false consciousness was abrogating for myself two key positions for the critical pedagogue: 'origin of what can be known and origin of what can be done' (Ellsworth, 1989, p. 323), placing myself beyond racism, sexism and all other oppressions (Gore, 1993).

The oppressive effects of my own emancipatory wishes also need re-exploration in the way I accounted for power relations between myself and teachers. I believed I could eradicate relations of power or equalize working relations through personal harmony and democratic action. All this was further underpinned by a view of adult learners who, if the proper conditions could only be established, would take responsibility for their own learning and become effective self-directed learners. Failure to do this meant my pathologizing teachers so that I could write at the time in my fieldnotes in 1987: 'Teachers are still not even recognizing that they are as oppressed as the pupils. They are still not even saying that they want to improve their teaching'. The deficit view this encapsulates seems so obvious now!

The point is that my own emancipatory discourse manifested itself in totalizing ways, imposing my own view of reality and appropriate practice so that teacher resistance was then interpreted as 'their problem' (McWilliam, 1992). Now, however, I see that the resistance lies rather in my own refusal to question such assumptions (or perhaps, vulnerability as a white university researcher working with african teachers in a deeply racist and hierarchical society). Teachers, after all, responded in terms of the positions available to them in discursive practices shaped by power–knowledge relations — what it meant to be a teacher in a Bantu education primary school, and how this constructed them as teacher-subjects in terms of who may speak about curriculum and who must be silent.

The asymmetrical relations between myself and teachers were nowhere more stark than in their early responses to the clinical 'gaze' of the facilitator-as-supervisor of their practice. What I originally saw as a problem or tension between being democratic and directive is now more nuanced for me. It was not so much a case of my not wanting to impose, as the fact that a face-to-face encounter 'is stacked in favour of the face that claims expert knowledge . . . the clinical gaze is that of the eye of power, while silent or self effacing regimes of power serve only to mask power' (Maurice, 1987, p. 246). Thus, the effects of my observing lessons were that teachers accepted their subordinate position, saw me as white university expert and provider of resources.

The point here is that disclaiming any imposition on my part and proclaiming a different positioning for myself did not make it so in practice where I was interacting with teachers with their own diverse biographies and their positioning within authoritarian discursive practices. I was trying to make teachers the objects of my emancipatory wishes, believing that emancipatory action research can stand outside power. But now I would argue that where knowledge (even critical and emancipatory forms) is being produced there are always also power and regimes of truth. The point is not to set out to empower or emancipate others, but to exercise one's own power in ways which enable others also to exercise power.

Counter Discourses

However, this also does not preclude power and control shifting from the margins to the centre, bringing to the fore subjugated knowledge, even where teachers still lacked the social power to realize their versions of knowledge institutionally. The question, then, is: What new positions might be made available to teachers through the development of a counter discourse of teaching which challenges the bureaucratic prescriptions of the education authorities?

I wish to illustrate this briefly by signalling the development of just one teacher, Veronica Khumalo,[7] a junior primary 'unqualified' teacher.[8] By the end of two years of our working together on reflective curriculum development in the teaching of reading, there was evidence that Veronica had taken up the position of producer-agent of curriculum knowledge in her own classroom, with colleagues, even with a school inspector:

And what I've discovered Melanie, we had a course with Miss Nama [school inspector] on the 4th [October]. What I've discovered, they were preaching there the group, the group teaching. And they explained this group teaching and it is said it has been introduced in the training colleges. And I said, 'Ah! We were doing this group teaching in Phakmisa with Melanie.' I said so to my principal. We even contributed there about this group teaching. You know what they said? They said you must group all the pupils and then you given different activities, different subjects. And I said, 'No! It won't work out. If you are doing English, you can't give another group a Maths activity. No, it won't work. That is confusion for the pupils.' [The others] they congratulated me! And the inspector herself said, 'No, let this be done like they are doing at Phakamisa'. (Interview 17, October 1989)

Moreover, Veronica now constructed herself not as a teacher working in isolation behind the private walls of her classroom, developing competence on her own, but in cooperative relations with others:

> I was just a self-centred somebody. I just go to my classroom, I teach, I go out, I go home. Now I discovered that, No! You must go to other people to other teachers. And you must also give help to other teachers . . . because I don't know everything, the other teachers know what I don't know. (Interview 17, October 1989)

All this was underpinned by her confident handling of the teaching process in her classroom (see Walker, 1991).

Change

For myself, then, what is also at stake in acknowledging power and its effects is not to be trapped by crippling pessimism, but to sharpen our self-awareness, and to see with greater humility and tentativeness what our efforts yield, recognizing that our knowledge of each other and the right course of action will always be partial, contingent and historically situated (Ellsworth, 1989).

The Politics of Identity

This has meant, in my own case, communicating across differences of 'race' to build fluid rather than inflexible, and above all respectful, cross-cultural relations. This was part of my knowing through action research, expressed this way by Veronica's school principal:

> There are some people, now who, I think you have outgrown now, you are not using your skin to communicate with teachers, some people use their skin, just because a person is white, and he's [sic] using that whiteness. (Interview with Mr Magona, 12 October 1989)

Moreover, I better understand now through theories of identity politics what it means to cross boundaries, which must not allow us to be paralysed by the pernicious legacy of racist attitudes in South Africa, or be seduced into a belief that only like can research like, thereby abandoning the political struggle over the concept of race. To accept this is as Miles (1989) argues, is to subscribe to a belief in some permanent essence of 'whiteness' or 'blackness' and a universal and unchanging truth about the nature of racism. Said reminds us that despite differences, identities, people, and cultures have always overlapped and influenced one another through 'crossing, incorporation, recollection, deliberate forgetfulness and of course conflict' (1993, p. 401).

But this means more than simply declaring oneself on the side of the oppressed or engaging in rhetorical moves which situate oneself, say as a white middle-class woman, while ignoring the socially constructed basis of such inequalities. A further related and important question here is whether a 'democratic' approach might leave us without critical resources for our own empowerment, given that not everyone has the same contribution to make or the same set of skills. It also does not mean allowing race to slip off the agenda in a 'non-racial' discourse. As a black British journalist recently pointed out in his coverage of elections in South Africa, apartheid and racism are not the same thing. Racism, he wrote, is a set of values, steeped in the history of colonialism and slavery, which oppress black people in whatever system they happen to live. So when apartheid goes, racism will still exist. What is at stake here for white democrats is confronting, however painfully, the dreadfully warped nature of apartheid society — we must face our race's identity and work even harder through 'friendship' (Ellsworth, 1989) to respect each other as women, as men, as black, as white.

From Common Sense to Good Sense

A further issue, and one particularly relevant for the 'new' South Africa, is that the tight connections between the work situation shaped by an illegitimate apartheid state — under-resourcing of black education at all levels, horribly overcrowded classes and lectures, and so on, and the way the work gets done through drill and practice, chanting, rote learning dull often incomprehensible lectures — needs be problematized. More than changing the work situation is then required, if, as Hatton argues:

> There is a significant group of teachers who do not seriously reflect upon ends and means in their work. For these teachers, changed working conditions are unlikely to change significantly the form of teachers' work. For these teachers, strategies formulated on the job have become teaching. (1988, p. 348)

I believe that *theoretically* informed action research is one way forward towards a different construction of teachers (in schools and universities) as flexible, critical and reflective practitioners able to develop quality education, and realize core values of equity and social justice. Such action research would challenge hegemonic voices or traditions which gloss over the different, multiple and competing voices subsumed broadly under the artificial unity of 'the people', 'the community' and 'the nation', and the suppression of self-criticism as fragmenting much needed unity which flows from this.

For me then, change, development and understanding through action research in the new South Africa lie in shifting from common sense to good sense, but in ways that recognize that the common does not exclude the good and the good can be fashioned out of the common. Cocks (1989), following Gramsci, explains:

> It is the porosity of commonness and goodness on which Gramsci counts when he asserts that the practical task of critical theory is to begin always with the commonsense and cultivate the elements of good sense studded through it, in order to fashion a new commonsense which is altogether good. (pp. 83–84)

From Personal Development to Collective Political Action

While we might begin with individual teachers in individual classrooms and lecture rooms, we cannot end there. While teachers I worked with demonstrated instances of agency, they were only instances and not yet a sustained narrative going beyond isolated moments of criticism. These moments did not then provide a new account of what it meant to be a teacher. Arguably, something more is required to turn gripes and complaints into a new account of what a teacher community is and means. Teachers worked to improve aspects of their classroom practice and to produce better lessons but this did not constitute curriculum transformation and could not without articulation with a political discourse and broader political struggles. Nor was authentic participation possible where teachers lacked the social and political power to define the terms and nature of their participation under structural conditions of inequality and hence asymmetrical project relations.

The question in the new South Africa is still, then, how do we progress from notions of personal development and reflection to more collective notions of political responsibility and social transformation? (Although making authentic connections between the micro and macro is still extraordinarily difficult, unless one understands political action as occurring at a number of different levels.) All this seems ever more complicated in South Africa where the old dichotomy between opposition to apartheid and complicity in its continuation generated a clear impetus to action for democratic educators. But the new complexities of reconstruction where it is now ever harder to find anyone (excluding the lunatic Right) who ever supported apartheid; where ethnic tensions as well as racial divisions must be confronted and understood; where Left political forces are for the moment marginalized; where gender needs more urgent attention in education, in politics, in the economy — all this may lead to ambivalence, even apathy. Again, Joan Cocks (1989) indicates the problem with admirable clarity:

> Can a militant oppositional effort be sparked by complex not simple ideas? Or is the power of simple ideas a necessary stimulant to rebellion?

> Is disillusionment then — when life ultimately is found to be compli-
> cated, not simple — rebellion's necessary end? (p. 84)

Cocks' question resonates also for me in relation to action research an earlier
plea for theoretically informed accounts. My point here is whether, when we
present action research as simple, accessible and always oppositional, we are
in danger of misrepresenting its complexity which will ultimately undercut its
potential for transformation?

I am also led anew to re-examining my own position — as a white,
female, middle-class administrator in a black university, working in management with black and white men, and a handful of women. Here I am drawn
increasingly to feminist theory to make sense of these experiences as I try to
fashion my own voice and struggle for institutional change under relations of
power shaped by gender, race and class. As Britzman (1991, p. 8) said of
teaching, and I would say of action research, it is 'always a process of becoming: a time of formation and transformation, of scrutiny into what one is doing,
and who one can become'.

Finally, then, under new conditions of possibility we must still contest the
purposes of action research: Whose problems do we try to understand? Who
speaks to and for whom? Who writes and who is written? Who, in the end,
benefits?

Acknowledgments

I want to thank three women for the conversations which helped me in re-
drafting this paper: Stella Clark, Sam Hollingsworth and Susan Noffke. My
special thanks to Bridget Somekh, friend, critic and colleague.

Notes

1 Lower case has been used deliberately to signal that 'african' is an indicator of an
  apartheid construction of a social or 'racial' group.
2 For purposes of analysis, South African universities are presently referred to as
  either historically white, that is those universities which up until recently have
  catered almost exclusively for students classified white; and historically black universities, all except Fort Hare, creations of the apartheid state to control and segregate black students (african, coloured and indian) in 'tribal colleges' called
  universities.
3 South Africa leads a list of thirty-six developing countries, followed by Jamaica, the
  Bahamas and Brazil, with Taiwan at the bottom as the least unequal.
4 In his capacity as Minister of Native Affairs, Verwoerd was responsible for the
  notorious Bantu Education Act of 1953. As Prime Minister of South Africa from 1958
  until his assassination in 1966, he is regarded as the architect of 'grand' apartheid
  and its attempts to segregate black and white at all levels; social, educational and
  political.

5 Fundamental pedagogics is an educational philosophy propagated by Afrikaner Nationalist academics to uphold 'Christian, national' principles, central to which is the assertion of segregation and white superiority disguised under a cloak of 'scientific objectivity'. Given Afrikaner control, in particular of teachers' colleges, fundamental pedagogics has been highy influential in the construction of teachers and teaching as an authoritarian process to mould children and lead them from ignorance, irresponsibility and incompetence to maturity. This is not to say, however, that it is the only discourse shaping teaching in schools. More recently, 'child centred' primary school practice has also penetrated black classrooms at a level of rhetoric, if less often of actual practice. See Walkerdine (1984) for a deconstruction of this particular discourse's regime of truth.

6 I want to thank Paula Ensor for drawing the concept of teaching as bricolage to my attention.

7 TMs is a pseudonym, as is the name of the school and its principal.

8 Since 1983, a 'qualified' teacher has been one with twelve years of school experience and a three-year post-matriculation teachers' diploma. Veronica had qualified under much earlier legislation according to which african teachers needed only ten years of schooling and a two-year teachers' diploma. The effect for Veronica and others in a similar position was that at the stroke of a pen they were labelled 'unqualified' with no state resources or support made available to them to upgrade their qualifications.

References

ADENDORFF, M. (1994) 'An evaluation of a post-graduate module in radical pedagogy'. Unpublished Masters dissertation, University of the Western Cape.

APPEL, S. (1991) 'Review Article: Changing your teaching: The challenge of the classroom', *Perspectives in Education*, **12**, pp. 103–6.

BRITZMAN, D. (1991) *Practice Makes Practice: A Critical Study of Learning to Teach*, Albany, State University of New York.

CHISHOLM, C. (1992) 'Policy and critique in South African educational research', *Transformation*, **18**, pp. 149–59.

COCKS, J. (1989) *The Oppositional Imagination: Feminism Critique and Theory*, London, Routledge.

COLYN, W. (1992) 'The schools and the classroom', in BREEN, C. and COOMBES, J. (eds) *Transformations? The First Years of the Mathematics Education Project*, School of Education, University of Cape Town.

DAVIDOFF, S., JULIE, C., MEERKOTTER, D. and ROBINSON, M. (1993) *Emancipatory Education and Action Research*, Pretoria, Human Sciences Research Council.

DOWLING, D. (1993) 'Mathematics, discourse and tokenism: A language for practice'. Paper delivered at the Second International Conference of the Political Dimensions of Mathematics, Johannesburg, April 1993.

ELLIOTT, J. (1991a) 'Action research and the National Curriculum: The art of creative conformity', in O'HANLON, C. (ed.) *Participatory Enquiry in Action*, Norwich, Classroom Action Research Network, University of East Anglia.

ELLIOTT, J. (1991b) *Action Research and Educational Change*, Buckingham, Open University Press.

ELLSWORTH, E. (1989) 'Why doesn't this feel empowering? Working through the repressive myths of critical pedagogy', *Harvard Educational Review*, **59**, pp. 297–324.

FOUCAULT, M. (1980) *Power/Knowledge: Selected Interviews and Other Writing, 1972–1977*, Hemel Hempstead, Harvester Wheatsheaf.

FOUCAULT, M. (1981) 'Questions of method: An interview with Michel Foucault', *Ideology and Consciousness*, **8**, pp. 118–28.

GORE, J. (1993) *The Struggle for Pedagogies: Critical and Feminist Discourses as Regimes of Truth*, New York, Routledge.

GRAMSCI, A. (1971) *Selections from Prison Notebooks*, London, Lawrence & Wishart.

HARSCH, E. (1980) *South Africa: White Rule, Black Revolt*, New York, Monad Press.

HATTON, E. (1988) 'Teachers' work as bricolage: Implications for teacher education', *British Journal of Sociology of Education*, **9**, pp. 337–57.

HORTON, M. and FREIRE, P. (1990) *We Make the Road by Walking: Conversations on Education and Social Change*, Philadelphia, Temple University Press.

KALLAWAY, P. and TSIBANI, F. (1993) 'School feeding as policy discourse: A priority issue in social reconstruction and educational planning for the new South Africa'. Paper presented at the third annual SACHES Conference, Salt Rock, October 1993.

KELLY, G. (1985) 'Action research: What is it and what can I do?' in BURGESS, R. (ed.) *Issues in Educational Research: Qualitative Methods*, London, Falmer Press.

LEIBOWITZ, B. (1991) 'Learning English as a first language in a multi-lingual classroom', *ELTIC Reporter*, **16** (1), pp. 16–24.

MAURICE, H.S. (1987) 'Critical supervision and power: Regimes of instructional management', in POPKEWITZ, T.S. (ed.) *Critical Studies in Teacher Education: Its Folklore, Theory and Practice*, Lewes, Falmer Press.

McWILLIAM, E. (1992) 'Towards advocacy: Post-positivist directions for progressive teacher educators', *British Journal of Sociology of Education*, **13**, pp. 3–17.

MILES, R. (1989) *Racism*, London, Routledge.

ROSE, B. and TUNMER, R. (1975) *Documents in South African Education*, Johannesburg, AD Donkery.

SAID, E. (1993) *Culture and Imperialism*, London, Chatto and Windus.

WALKER, M. (1991) 'Reflective practitioners: A case study in facilitating teacher development in four African primary schools in Cape Town'. Unpublished doctoral dissertation, University of Cape Town.

WALKERDINE, V. (1984) 'Developmental psychology and child-centred pedagogy: The insertion of Piaget into early education', in HENRIQUES, J., HOLLWAY, W., URWIN, C., VENN, C. and WALKERDINE, V. (eds) *Changing the Subject*, London, Methuen.

WEINER, G. (1989) 'Professional self-knowledge versus social justice: A critical analysis of the teacher-researcher movement', *British Educational Research Journal*, **15**, pp. 41–51.

WINTER, R. (1987) *Action-research and the Nature of Social Enquiry: Professional Innovation and Educational Work*, Aldershot, Avebury.

WOLPE, H. (1991) 'Some theses on people's education', *Perspectives in Education*, **2** (12), pp. 77–83.

4 Changing the Culture of Teaching and Learning: Implications for Action Research

Peter Posch

ABSTRACT *Action research has become an international movement in a relatively short time. In this paper I want to show that action research is a correlate to changes in the culture of teaching and learning which in themselves are answers to global changes in industrialized societies. I shall first describe two generic social developments: increasing complexity and growing individualization. They produce some of the basic challenges for the future of education which are described in the second section. Third, I want to sketch and illustrate some of the elements of a developing educational culture. In this framework I want to show that action research is a necessary development to assist teachers and schools in coping with dynamic developments, divergent demands and complex practical situations.*

Two Developments in Industrial Societies

The Complexity of Conditions of Life

One could say that the increasing complexity of brain functions allowed human beings to step out of animal life by becoming conscious of their ability to influence action and to gradually increase control over instincts. J.G. Herder, a German philosopher, created the metaphor of the 'the first freedman of creation' (der erste freigelassene der schopfung).

In our millennium, in fact in the last three hundred years, the human capacity to access, store, process, use and communicate information has surrounded humans with a rapidly increasing array of artificial 'cognitive limbs' and has created a new type of complexity; a complexity not only within each person (in terms of brain potentials), but also and perhaps primarily between people and their human-made environment, i.e., their artificial extensions of their anatomical outfit. The huge variety of artificial cognitive limbs and their manifold interactions with natural systems have created the complexity which we experience today. The effects and side-effects of this development have become less and less foreseeable and controllable and have created a situation in which the enormous 'life potential' of economic and technological development is confronted with a growing 'death potential' (i.e., through global threats created by this development).

But this is not the only effect. Another significant effect of growing complexity is the decrease of the problem-solving capacity of large socio-economic systems and centralized power structures. This is illustrated in tendencies to 'privatize' public services, to devolve responsibility for environmental quality, for social security, and even for safety to the individual citizens. In the past, the ability and willingness to shape professional and public life has been limited to the elites in society, to politicians, entrepreneurs, academics, artists. Now, it seems that these competences and the values associated with them are demanded from more and more citizens. Let me give you three simple illustrations from the emerging patterns of work:

- The proportion of human routine activities, i.e., of those activities for which the correct algorithm can be predetermined, is decreasing. They are being taken over by technical systems. As a result, the demands on professional qualification increase.
- The mutual dependence between hierarchical levels increases. As a result, cooperation in teams gains in importance.
- The heterogeneity of demands on the individual employee increases: more and more they comprise organizational, executive and supervisory tasks.

The Individualization Process

Complexity and growing risks have been enforcing a decentralization of initiative, responsibility and competences. One can assume that elements of power are only given up if, through a given distribution of power, it is no longer possible to keep the diversity of influences in society under control. So power is devolved inasmuch as this appears to be necessary in order to sustain power. The paradoxical character of this relationship indicates that this is not a linear process. However, this process has enormously increased the interdependence of people. Paradoxically again, this interdependence can be seen as a primary force towards individualization. The increasing diversity of demands provides more and more individuals with influence over others and through this process their potential 'value' increases.

There are phases in the devolution of power; at first, power is devolved to subcultures or lobbies if the centre is no longer able to control the emerging diversity. As more and more subcultures develop and/or the diversity within subcultures increases, the ability of the power structure to provide safety and legitimization within this system again decreases. The social units that have to develop their own coping and adaptation strategies become smaller and smaller moving towards the smallest possible social unit, the individual.

There is little dispute about the global process of individualization. For Sloterdijk (1993, p. 48) this is a late stage of human development, when 'cosmic singles' are populating bigger and bigger cities. According to Fend (1990, p. 50), two hundred years after the Age of Enlightenment, 'the right, duty and possibility to use one's mind without being led by somebody else, and to shape one's life at one's own terms is only now becoming a widely held claim and an emerging reality' (or in Frank Sinatra's words: 'to do it my way').

This individualization process highlights a serious problem that has been described by game theory. Let me illustrate it with an example. Two players are involved in the same game. Let's assume that they are playing against a bank. Both have two strategies at their disposal: to defect and to cooperate. If both of them cooperate, they gain three points each for cooperation. If both defect, their gain is one point each. If one deserts and the other cooperates, the defector gets five points and the other nothing. What would be rational behaviour in such a situation? The rational player would clearly choose to defect. This is the best choice if the partner cooperates. The gain is five points. But it is also the best choice if the partner defects. The gain is still one point and cooperation would bring nothing. Therefore both players will defect and receive one point. If both had cooperated each would have gained three points (Sigmund, 1993, p. 46).

This rather formal model is a variation of the prisoner's dilemma. It is not too difficult to illustrate by examples from everyday life. It may explain a widespread and puzzling attitude towards the environment. One's own environmentally sensitive ('cooperative') behaviour brings less comfort and has in many cases no advantage if all others do not care (i.e., if they 'defect'). On the other hand, one's own carelessness is very advantageous if the others behave in an environmentally sound way. As a result, it is 'economical' to do both: to plead for an environmentally responsible behaviour of others and not to care about it in one's own behaviour. There is an interesting message in this example. If we start from the assumption that, in general, people do what they consider useful to them, one would expect defective rather than cooperative (i.e., morally defensible) behaviour. And this is actually what we get.

Challenges Schools will Face in the Future

Using these two sketches of ecological and social developments, what are some of the challenges schools will face in the future and which will demand answers in terms of changes in the culture of teaching and learning?

Challenge No. 1: Negotiation of Rules

No society can survive without rules and conventions. If established structures lose some of their stability and legitimation, the rules have to be developed by negotiation. This process has started in many areas of life. An important example is provided by the socialization of the young in families. The authority relationships in families have dramatically changed during the past thirty or so years: what is allowed and not allowed is no longer a lone parental decision but is negotiated. With these experiences pupils come to school and are confronted with a culture of predefined demands and without space for negotiation. This clash of two 'cultures' appears to be the reason for many conflicts. Schools in general have not found ways to cope with a social development in which negotiation of rules and norms is gaining importance. How can they contribute to this process.

Challenge No. 2: Social Continuity

If traditional social networks lose stability and legitimation, social continuity and social control are reduced. However, continuity of social relationships appears to be an indispensable condition for cooperative behaviour and social responsibility. This has also been illustrated by experiments in game theory. If a relationship is kept up for a long time, tendency to defect decreases and the mutual trust that is necessary for cooperation can develop (Sigmund, 1993). It is perceivable that one of the reasons for the apparently increasing violence in big cities has to do with the decline of continuity in society, which has accompanied the individualization process. It has promoted the tendency to instrumentalize social relations in terms of short-term gains: to defect and not to cooperate. Can schools create situations in which the young experience continuity in social relationships and are shown that to cooperate is better than to defect?

Challenge No. 3: Dynamic Qualities

The growing complexity of public, economic and private situations enforces decentralization of decision-making structures. This means that more and more individuals will have to be able to cope with unstructured situations, define problems, take positions and accept responsibility for them. As an example, the power of political authorities to change basic conditions of life, such as the production and consumption of energy, is increasingly dependent on individual initiatives in the population. Presently there is an incongruity between the diversity of influences young people feel subjected to and their own opportunity to exert influence. There appears to be a growing quest among the young to be taken seriously, to be able to influence their conditions of life and to leave traces in their environment. A quotation from a seemingly well-educated skinhead in a radio report gives a vivid illustration of this hypothesis: 'I throw stones, ergo sum' is an interesting rephrasing of Descartes' statement on the nature of human existence, 'I think therefore I am'. There is a need for frameworks in which the young can contribute to shaping their environment in a responsible and constructive way and experience that they 'matter' in society.

These arguments provide a strong case for the promotion of dynamic qualities in the wider population. Can schools provide opportunities for the young to experience that they can make a difference?

Challenge No. 4: Reflection and a Critical Approach to Knowledge

Beck (1986, p. 35) has argued that the understanding of most risks in industrialized societies is transmitted by arguments. Consciousness of risks *is* a form of theoretical or 'scientific' consciousness, because causal interpretations of events cannot be observed and are therefore theoretical. As a result, the dependence on scientific knowledge has enormously increased; on the other hand, however, science is regarded also as a producer of risks and therefore tends to be met with increasing doubt and scepticism. This creates a paradoxical situation in which substantive knowledge becomes as important as a fundamental scepticism against whatever is offered as knowledge. It may imply that the definition of 'important knowledge' can no longer be left to traditional authorities only but must be established also by negotiation involving more and more peripheral and smaller social units. As a result, communication between these social units becomes important as different perspectives have to be dealt with and become a basis for establishing 'local truth'. 'Confidence shifts from confidence in contents — established rules, data, methods — to confidence in processes that allow us not to eliminate but to keep error under control' (Losito and Mayer, 1993, p. 72). Can schools provide opportunities to combine advocacy of knowledge with enquiry and to promote both an appreciative and critical stance towards knowledge?

Elements of a Culture of Teaching and Learning for the Future

The prevalent cultures of teaching and learning are still attuned to a relatively static society, in which the necessary knowledge, competences and values are predefined and stored in curricula, tests and accredited textbooks. Schools are expected to prepare the majority of children and young people to meet satisfactorily the demands others have defined for them. The main characteristics of this culture are as follows:

- *A predominance of systematic knowledge.* Priority is given to well-established facts, allowing schools to maintain a close relationship with the results of academic knowledge production. Low priority is given to open and controversial areas of knowledge and to personal experience and involvement.

- *Specialization.* Knowledge is compartmentalized in subject-matter fields which more or less correspond to the academic disciplines. This again facilitates an orientation towards established standards of quality and gives teaching and learning a clear and predictable structure. On the other hand, complex, real-life situations tend to be disregarded because they cross the disciplinary boundaries.

- *A transmission mode of teaching.* This mode facilitates the retention of the systematic character of knowledge and its reconstruction by the student. It tends to discourage the generation and reflective handling of knowledge.

- *A prevalence of top–down communication.* This facilitates the external control of pre-defined knowledge structures, provides stable frame conditions, and facilitates the maintenance of control in the classroom. However, it discourages self-control and cooperation among students (or teachers) and networking within and across school boundaries.

These few arguments indicate an important and difficult feature of the culture of teaching and learning in the future. This culture will have to comprise contraries. It will have to retain the strengths of static elements and we have to complement them with dynamic ones. The balance of the two will have to shift if schools attempt to find answers to the social changes presently occurring. What are some of the dimensions of a learning culture of the future? Let me give you a few examples which are still compatible with present infrastructural possibilities.

In a school in Austria, one day a week in a class of 11th grade students was declared as project day. Three subjects provided the necessary curricular time: chemistry (one hour), social science and religious education (2 hours each). The issues and patterns of work for each of these days were decided upon by a steering group of students of this class together with the teachers of the three subjects. Some of the issues identified were: communication, a child is born, the environment, science and research, being handicapped, etc. The procedures consisted of visits, e.g., to the university or to a home for the handicapped, of invited lectures, readings, workshops prepared by students of the steering group for their colleagues, etc. For each two or three issues, a newspaper was produced in which major findings, experiences and reflections were documented. Through the year each student of the class participated at least once in the planning activities of the steering group. The basic idea of the whole exercise was to gain understanding and substantial knowledge on salient themes as perceived by the students (Rauscher, 1986). In this example students are confronted with situations which are not entirely restructured but have open spaces. They get the opportunity to decide certain things on their own. Work on this project day is not assessed by the teachers for grades. Quality control is provided by the students themselves in their reflections on the process.

In a school in Scotland a student group carried out an environmental audit in their own school to find out about strengths and weaknesses and provide a basis for action. In another Scottish school, students were involved in several investigations on Travel, Tourism, Transportation and Tipping and results were made public through exhibitions and weekly articles in a local newspaper (McAndrew and Pascoe, 1993, pp. 23, 33). In both examples 'local knowledge' was produced which had not been available so far and which was considered useful for a specific audience.

A final example shows still another dimension (Axelsson, 1993): all teachers and students of Pärydskolan in Sweden have taken up challenges in their own vicinity and reserved two hours per week for environmental projects. They started from their own interests and from requests from outside. For example, a parent had ponds that were now overgrown with trees and plants — could the school help them with these? The school now has fish breeding in the ponds. Similar initiatives resulted in a shop in which the school sells environmentally friendly detergents; in a green house; and in the restoration of a water mill. Some of these initiatives involved hard work: if it turned out too hard for students to do themselves, parents came to the school on Saturdays to help.

> The school has managed . . . to become the centre in the small village.
> People come to school asking for help about ponds, about acid water
> in their wells and about different environmental issues. The school is
> a place where knowledge is to be found. (Axelsson, 1993, p. 42)

In this example an additional dimension is illustrated. Students are able to leave traces in their environment. They can participate in shaping some of the conditions in which they live and experience that what they do is not only aimed at competences to be utilized in the future but is making a difference here and now.

In these examples the concept of learning is extended: the dominant paradigm of learning is based on the classical separation of knowledge and action and on the hope that transmission of knowledge will enable students to act responsibly in the future. An extended concept of learning views learning also as a process of joint seeking, joint experimenting and joint construction of reality. On a more fundamental level, it affects the structure of exchange processes at the heart of education (*cf.* Elliott and Rice, 1990). As a 15-year-old student put it: 'To prepare for life means to do something now' (Mair, 1990, p. 9).

What Does this Mean for Action Research?

Teachers and students who leave the 'stable state' (Schön, 1987) of systematic knowledge transmission and involve themselves in school initiatives of this kind have to cope with open-ended, uncertain, unpredictable, sometimes contradictory situations entailing risks. In situations of this kind, teachers are often unable to control the teaching and learning process alone but have to rely on their students' capacity for self-organization. As a consequence, an interest in negotiating work with students develops. Often, competences in other subject-matter fields are needed. As a result, an interest in communication and interdisciplinary cooperation develops. If teachers involve themselves in initiatives of this kind, they become highly dependent on the flexibility of infrastructural conditions for their work — such as curriculum, time budget, use of space and school facilities, materials, etc. — so an interest emerges in influencing the regulations determining the structures of work in school. Finally, in many cases, these teachers need the cooperation of persons and institutions outside school. As a result, an interest develops in building networks of communication with other parties and to cross the boundaries between school and environment (Posch, 1994).

These types of interests arising from characteristics of the teachers' work are likely to be a breeding ground for action research. In these contexts, systematic individual and collaborative reflection on action and communication about the knowledge that has been generated appears to be a 'natural' means to realize the values of teachers and students (Altrichter, Posch and Somekh, 1993). At present, action research is in most cases an element of externally led courses promoting teachers' career interests or an ingredient of externally stimulated research projects. If schools move towards a more dynamic culture of teaching and learning, action research will become less dependent on these external structures but will more and more become a necessary correlate of what teachers and students want to do.

It is, I believe, not a coincidence that Stenhouse's idea of the 'teacher as researcher' developed in the framework of the Humanities Curriculum Project (Stenhouse, 1975, p. 49). In a curriculum based on the discussion of controversial issues with the teacher as neutral chairperson, open spaces are created and teacher control is reduced. In such a context 'reflective conversations' (Schön, 1987) with lowly structured situations become imperative not only to keep risks under control but also continually to create understanding as a basis for action. In the meantime the social pressures on schools to allow students to experience and to create meaning have increased — even if they are not in line with present governmental policy as it appears to be the case in England. As a result, Stenhouse's view of 'extended professionalism' becomes even more relevant than it was in his time: 'The concern to question and to test theory in practice as a basis for development' (Stenhouse, 1975, p. 144). These challenges are probably the main reasons why action research is becoming an international movement.

Traditionally, schools are the recipients of demands from power structures in society. In the future it will be necessary for students and teachers also to express and realize their views of the society in which they want to live. Action research is in a sense only another word for this.

References

ALTRICHTER, H., POSCH, P. and SOMEKH, B. (1993) *Teachers Investigate their Work: An Introduction to the Methods of Action Research*, London, Routledge.

AXELSSON, H. (1993) *Environment and School Initiatives — ENSI*, Goteborg, University of Goteborg Department of Education and Educational Research, Report No. 1993–01.

BECK, U. (1986) *Rtstkogese Uschaft — Auf dem Weg in eine andere Modeme*, Frankfurt, Suhrkamp.

ELLIOTT, J. (1991) 'Environmental education in Europe: Innovation, marginalisation or assimilation', in OECD/CERI (ed.) *Environment, Schools and Active Learning*, Paris, OECD/CERI, pp. 19–39.

ELLIOTT, J. and RICE, J. (1990) 'The relationship between disciplinary knowledge and situational understanding in the development of environmental awareness', in PETERS, M. (ed.) *Teaching for Sustainable Development*, Report on a Workshop at Veldhoven-Netherlands, 23rd–25th April 1990, Enschede, Institute for Curriculum Development, pp. 66–72.

FEND, H. (1990) 'Btldungskonzepte und lebensfelder jugendlicher im sozialmstoiischen wandel', in LETTNER, L. (ed.) *Ite Offnet sich die Schule Neuen Entwicktungen und Aufgaben?* Wien, Bundesverlag, pp. 42–66.

LOSITO, B. and MAYER, M. (1993) *Environmental Education and Educational Innovation: Italian National Report on ENSI Research*, Frascati, Centro Europeo del Educazione.

MAIR, G. (1990) 'Lernen durch Handeln in Projekten: Schtlicrinnen beeinflussen die kommunale Umweltpolitik'. Bundewwbeitsgemetmchaft Bildung und Die Grunen (ed.) Forum zur Okologischen Bildung vom 14–15 September 1990 in Mimberg, Berlin, BAG Blidung/Die Grunen, pp. 8–10.

McANDREW, C. and PASCOE, I.P. (1993) *Environment and School Initiatives (ENSI) Project in Scotland, The National Report: Case Studies in Environmen Education*, Dundee, Scottish Consultative Council on the Curriculum.

POSCH, P. (1994) 'Networking in environmental education', in PETTIGREW, M. and SOMEKH, B. (eds) *Evaluation and Innovation in Environmental Education*, Paris, OECD/CERI.

RAUSCHER, E. (1986) PRODO — 'Ein projektorientlerter unterricht stellt sich vor', *Erziehung und Unterricht*, 4, pp. 242–9.

SCHÖN, D.A. (1987) *Educating the Reflective Practitioner: How Professionals Think in Action*, New York, Basic Books.

SIGMUND, K. (1993) 'Spiel und biologie: Vom mitspielen zur zusammenarbeft', in RELCHEL, H.-C. and PRAT DE LA RIBA, E. (eds) *Naturwissenschaft und Welbild: Mathematik und Quantenphysik in unserem Denk- und Wertesystem*, Wien, Holder-Pichler-Tempsky, pp. 45–58.

SLOTERDIJK, P. (1993) *Im selben Boot*, Frankfurt, Suhrkamp.

STENHOUSE, L. (1975) *An Introduction to Curriculum Research and Development*, London, Heinemann.

THONHAUSER, J., MOOSBRUGGER, M. and RAUCH, F. (1994) 'Evaluation of the Austrian Environment and School Iniatives Project'. Research Report commissioned by the Institute for Interdisciplinary Research and Continuing Education, Salzburg.

Part II

Action Research and the Development of Contexts

5 Is there a Difference between Action Research and Quality Development? Within or Beyond the Constraints?

Christine O'Hanlon

What is Quality Development?

The concept of quality is a recurrent theme in the political rhetoric of educational policy. It is a significant new key word in the contemporary political process of creating meaning and purpose in education. It is endowing schools and colleges involved in learning and teaching, and the assessment of student learning with desirable values which are embodied in prescribing and evaluating educational processes and outcomes. Notions of quality vary according to how the various parties concerned with it, i.e., politicians, administrators, teachers and parents, compete for dominance and use the 'Q' key word in policy, planning and practice.

Quality is a specific metaphor which has evolved from industrial and commercial contexts. The quality revolution in the world of industry and commerce is based on a re-evaluation of how people are managed at work, recognizing that the greatest single resource in any business enterprise is its workforce. The rationale for using the Quality Development (QD) approach in education has resulted from government pressure for schools to operate more as successful business enterprises. However, the key question is — does it benefit the teaching profession to use theories about best practice in industry and commerce? Can 'service industries' like education benefit from commercial rhetoric and practice? QD practice in education has used ideas and practices successful in industry openly from the outset. Much of the QD business, terminology and language and its underlying precepts have been directly transferred to education. Examples are:

- Quality Development itself based on the key word 'quality' based on the work of quality gurus in industry e.g., Crosby '70, 84; Deming '82, 86; Ishikawa '76, 85; Juran '79, 80; Oakland '86; Peters '82; and Taguchi '79, to name just a few.
- The concept of 'building quality in' as opposed to external inspection.

- The importance of valuing the experience and expertise of the workforce (teachers).
- Involving the customers (pupils/students, parents and employers).
- The value of defining strategies for achieving the aims omitted in the quality gap i.e., target setting and action planning.
- Definition and agreement on expected standards, measurement of performance and for success.

There have been a number of QD initiatives in education in recent years that have involved large numbers of educational professionals in the monitoring and examination of their practice, yet they follow a model not unlike the general principles that are recognized as key factors in action research. A key example of the QD process to which I will refer to explicate my argument are 'Staff Development Resource' packs published by the City of Birmingham Education Department. The first one I refer to is entitled *Developing Quality Through TVE (Technical and Vocational Education)* which was published in 1990. It was aimed at developing a quality service using the principles of supported self-evaluation. The resource pack is a very user-friendly, step-by-step approach to educating professionals about QD. Is this the same as improving practice as used by proponents of action research? I want to examine this issue further as there are some important principles that are integral to these observations. In this document it is stated that: 'A high quality service is one which is both efficient and effective in meeting client needs.' The language used is very much 'management' language. The pack is concerned with providing a high quality education for pupils and students, satisfying the needs of other groups concerned with education and training, and TVE is used to help satisfy these needs.

Evaluation is defined in the pack as 'a systematic process of making judgments about quality' involving the collection, the analysis, and the interpretation of information. On the same page we have the phrase 'Self-evaluation generates improvement and development.' The main premise for the activities in the pack is that quality can be developed through evaluation, which requires *Improving, Proving and Learning:*

Improving is defined as informal, formative and makes 'service' better.

Proving is concerned with accountability, and is summative and formal.

Learning is sited in the methods of action research and is concerned with highlighting the needs of staff, curriculum and institutional development.

According to the advice given in the pack one begins the evaluation process by collecting information about the current situation, then one is asked to identify a 'quality gap' between 'where you are now' and 'where you would like to be'. (This is not stipulated in terms of the person but more the school perspective.) From the quality gap comes the *issues* of concern, the *criteria* from established external institutions and bodies, *personal judgment* and action. The action plan emerges from the judgments made about the situation. Collaboration with a colleague is encouraged on the action planning, entailing a review of roles and responsibilities, opportunities and constraints, possible solutions, resource implications, choice, outcomes and targets. Evaluation is recommended through the process of interpretation and analysis, through keeping the work manageable, examining reasons for the evaluation, considering the audience, deciding on persons to be involved and identifying appropriate methods. It is unclear whether the methods are the research methods or the curriculum improvement methods.

So far there is a lot of consonance with action research. Ideas about analysing the information are given as well as its recording and reporting. The learning style of Kolb is referred to as a rationale for experiential learning when the concept of reflection is introduced with a cycle of experience-reflection-conceptualization-experimentation. Reflection is used in the pack as synonymous with thinking, it isn't fully explicated, and reflection is seen to be reflection *on* learning experiences.

The most recent Quality Development pack to emerge from Birmingham LEA is a much larger pack with the aim of facilitating change and development in schools. It is entitled the *Quality Development Resource Pack* (1993) and has evolved from the previous work on quality which was targeted at further education colleges in 1985. The introduction to the pack expresses the promise that: 'QD enables teachers in schools to work together to improve the quality of learning and teaching.'

This aim mirrors in many ways the aims of the newly named Collaborative Action Research Network, which has been set up to develop improvements in teaching and learning in schools through action research defined as 'the study of a social situation with a view to improving the quality of action within it' (Elliott, 1989). Although the QD pack openly uses action research as a basis for training the other three sources for the rationale are acknowledged from:

1. Educational evaluation literature.
2. School effectiveness and school improvement literature.
3. Quality in industry and commerce.

The different sources form a tension in the language and expression of ideas in the pack. The literature on educational evaluation is the source of the proving idea based on the relationship between accountability and professionalism (Simons, 1987; Holly, 1986) and the improving idea comes from school effectiveness literature (Fullan, 1992; Reynolds, 1992; Hopkins and Ainscow, 1993). The action research element is largely related to Stenhouse's ideas for curriculum improvement and Elliott's (1991) work on action research. The school effectiveness literature is based on questions related to the extent and nature of school effects and the in-school factors associated with them. The concept of quality in industry and commerce which was influential in the first QD pack is re-introduced and explained by the addition of the terms — quality gap, customers, mission, performance measures, the learning organization, leadership — and the importance of the involvement of senior management is emphasized in change and innovation processes.

There is ample evidence provided for the bringing together of action research and self-evaluation for school development planning, appraisal, Local Management of Schools (LMS) monitoring, assessment monitoring, including Standard Assessment Targets (SATs), and a range of Local Education Authority (LEA) reviews required to help prepare schools for inspection. The action research component of the training provides the process and the key elements for teachers to put QD into practice in schools. It is also intended to be an element of whole school QD policy and to become a regular feature of school life, particularly for those priorities in the School Development Plan that need an indepth 'research' mode of development.

Criteria that one uses to make judgments within the research process are related to school-based success and effectiveness which focus on the question — *What differences do schools make to pupil learning?* Success is related to the question — How has the action research project (carried out under the supervision and direction of the university) affected the quality of education at your school? The research process is presented as being both challenging and rewarding. Challenging to the teacher because it is necessary for him/her to identify specific criteria to become more effective. Rewarding because once the success criteria are decided upon they then provide an anchor point for the whole project. The challenge and reward provide both a device for focusing and a yardstick against which to measure data and to sustain teachers' interest in school development through QD.

A Quality Development Training Programme for Managers

I feel a close relationship to the whole process of QD because after many years of using action research as a basis for professional development in schools, I found myself involved as the director in an ambitious QD training programme with Training and Enterprise Council (TEC) staff in management roles. During this time I used the Quality Development Resource Pack as a framework for the initiative. However, because I was basically an experienced action research tutor I experienced some tensions and dilemmas in using the pack. Some of these I want to expose now in a broad critique of the QD approach to professional development.

In the QD programme that I directed there was a total implicit focus on the customer and the organization's needs, subsequently it was the aim of the participants to produce an action plan to benefit them, before all else. The managers saw the action plan as providing the security and the framework for the proposed evaluation or action which was terminal, i.e., there was an end in sight and it was seen to be a finite plan to satisfy the organization's needs.

In the training programme, we established an early sense of purpose and proposed aims for the staff's action, which was quickly secured in order to avoid the discomfort and uncertainty of unresolved direction. (This would, it was implied, be wasting valuable training time.) As the programme director I attempted to use an action research approach to the training programme which I believed at that time was consistent with the QD approach. After a process of defining the participants' issues and concerns I reviewed all the issues for the groups and discussed what was possible to research or evaluate by dividing the issues into 'insider' and 'outsider' issues. The main themes that emerged came from 'outsider' concerns which I felt needed to be reformulated as 'insider' problems before we could consider individually the nature of what would be effective action. The issues of government policy, the recession, and unemployment were changed with some effort and after further enforced reflection and discussion to:

- an understanding of and a commitment to quality development in the organization;
- improving networking and relationships and employer interest;
- leadership, accreditation;
- customer needs;
- assessment;
- resources;
- finance;
- staff training.

I kept a record of what took place during the twelve-week programme and find my notes from an early group planning meeting point to the repetition of pleas for 'action planning as soon as possible' and 'the challenge is moving things forward faster', regularly expressed by team members who formed a specific focus group for the project duration.

There was an implicit pressure on all participants for action planning, which is a well-worn and well-known management strategy in business and industry — *as soon as possible.* As a result of which it was difficult, if not impossible because of the time constraints in the training programme, to take the time I thought necessary to reflect on the present situation and consider carefully the implications of the 'audit' or 'reconnaissance' which had begun as a review of the participants' current situation.

The Tension

Instant judgment was seen to be a strength of participants because they were in management roles. The tension I experienced was working *against* a culture which valued immediate responses and judgment, which would in the future progress of the course also relate to the process of 'evaluation'. It was seen to be an intellectual strength to make an instantaneous decision to select and assess quickly what was or was not of significance to include in the action plan. Deliberation was viewed as prevarication and action was just 'doing' it. Every sensible, experienced manager knew instinctively, because of their experience, what needed to be done and they were happy to pass their wisdom on to other participants and to tell them 'what' to do and 'how' to do it: a situation which I tried to counter in group exercises and workshops. I was confronting an instantaneous action culture which I believed required to focus on more analysis of the present problems and further dialogue about possible future action. This instantaneous view was sustained and perpetuated by the managers because it appeared that the management culture could not pragmatically entertain any alternative framework, which then resulted in the course director being seen as deliberately provocative, slow and deviant. This perception I believe came from my attempts to engender more reflective practice and a more self-centred subjective focus to the research issues.

The short university input into the QD training was expected to get results, and this began with the action plan which was developed in a practical way to further the institutional aims of the TEC. There was a final product in the form of an account of the evaluative process and the monitoring of the action which resulted from the implementation of the 'action plan'. The course participants reviewed their concrete data which was predominantly in the form of facts and figures, thereby avoiding their personal contributions to their work and any professional self-evaluation. It is one of the gurus of QD in business and industry, Tom Peters, who states: '*If something cannot be measured it cannot be improved*'. Most participants took this phrase literally in their planning with the result that concrete measurable data was much in evidence. The emphasis was very much upon the product rather than the process of self-evaluation or evaluating the quality of the action. The training course became a team-building exercise in collaborative 'action planning'. Everyone involved in it felt good about that, but why didn't I?

Another problem with the management training initiative was the rhetoric and language used by participants. The language of the pack was a language borrowed from the business world. Words such as 'targets', 'stakeholders', 'mission statement', 'audit', 'tasks', 'clients', 'efficiency', 'products' and 'service' were all foreign to me in a reflective educational process model and only indicated to me their derivation from an unfamiliar cultural sphere. Using business terminology in an educational process I found difficult to accept fully because I felt that the business language and rhetoric was deprofessionalizing the professional within a specific educational process. From the business perspective it is the 'customers' who measure the quality of both the products and the service. This principle cannot simply be translated to educational contexts to be applied to parents and students. In an educational process the teacher is a facilitator of pupils'/students' learning processes and s/he is not a technical operative of educational ends as the managers, in this instance, considered themselves to be.

The two main quality tasks in the industrial metaphor are Quality Assurance (QA) referred to as 'All activities and functions concerned with the attainment of quality'; and Quality Control (QC) which is 'The operational techniques and activities that sustain the product or service to specified requirements'. To guarantee that continuous improvement of the service is maintained TQM is invoked in the QD process models to aid school inspection. TQM or Total Quality Management is an industrial and commercial metaphor which is used to ensure that all personnel in an organization know what quality means and that all their activities are performed in a quality manner. Typical of a quality organization would be:

- the involvement of everyone
- quality improvement which is systematic and continuous
- more concentration on customers
- everyone must know competitor quality
- education and training is part of the process
- quality policy is public and regularly reported. (Themes reflected in the QD process.)

It is not difficult to see a parallel process in these aims which, when applied to schools, would not but support new and innovatory practice. However, it is in the interpretation of the phrases that divergences arise between QD as a technical operation and QD as an educational process.

The argument in this paper is that there are similarities and differences between action research and QD in professional educational contexts. The QD approach has much to recommend it as a means of empowerment for teachers in schools, to enable them to experience the limits of their power and control of (self-chosen) initiatives in individual schools and institutions. However, because it is based on the management metaphor and on the implicit implementation of government policies (which are not self-chosen), it is inevitably going to give rise to criticism. There is no resistance to policy directives hinted at in the QD process. There is an encouragement to change and improve school practice but the interpretation of this may be a technicization of curriculum products through improving already accepted and current aims and methods.

Another separation between the processes resides in the QD explication of self-evaluation. It is the school or the institution in QD which does the self-evaluation: 'School — Know thyself!' The school undertakes QD through the QD coordinator who is responsible for moving QD forward in the school:

- for the facilitation of the consultative and decision-making process with appropriate groups;

- to work collaboratively with colleagues to represent the school at LEA level;
- to act as liaison between outsider and insider colleagues and to prepare QD documentation.

Therefore the assumption is that the QD coordinator supports the school's aims and the aims of QD, without challenging its underlying assumptions.

However any new proposals for improving the quality of teaching and learning needs to take into account the existing professional culture from a critical perspective. The pack begins with the assertion that the QD approach should be built into routine working practice: be conducted collectively, systematically and rigorously; be enacted in such a way that the processes support development and involve all who have a part to play including (where appropriate) pupils and students. The pack stresses throughout, the necessity of QD being a collaborative venture, entered into by the whole school and includes a copy of an action research contract signed by both the headteacher and the QD coordinator who is the key person in initiating the QD process in the school. QD is seen as a shared enterprise which in the present climate of rapid change in schools is to be welcomed as a democratic process supporting teacher empowerment. This appears to support the same principles as a collaborative action research process.

The Role of Higher Education in Quality Development

One of the strengths built-in to maintain QD in schools is the provision of an academic award — an Advanced Certificate in Education (ACE). The LEA supports teachers in obtaining the award to promote the QD principles in local schools. The opportunity is therefore provided for academic intervention and the facilitation of the QD process. The question that arises from this partnership is: What should the role of higher education (HE) be in the QD process? The principles of the process are already set out in the QD pack, so is it simply a matter of HE providing facilitation, direction and depth to the process? The participants in the QD process need critical feedback from outside the school and a critique from within the school based analysis of the research concerns facilitated by their action research. The HE tutor as a facilitator of the action research process, shares with the teachers in the role of transforming schools. This role challenges the HE tutors to become 'insider researchers' in some respects although their role is very much one of an 'outsider'. It is clear that the teachers are the 'insider' researchers in their schools which makes them the agents of change in the QD or action research process. However, the QD facilitator is an 'outsider' in terms of 'knowing' about the school situation in terms of acquiring information and 'knowing' about where the quality gap exists between 'knowing where we are now' and deciding 'where we want to be'! The HE tutor is an 'outsider' looking inside through the QD process. S/he does not have to react quickly in the heat of the participant's action and make quick decisions about what needs to be done in the school 'now'. This places them in an advantageous and less active role in the relationship.

Ideally the facilitator's academic role is to provide the space for teachers' personal reflection on the situation in an intellectually challenging and critical context. To prevent teachers and managers from rushing to settle ambiguities in the professional situation and taking action 'now', the principles of Socratic teaching may be employed to encourage reflection about alternative perspectives in action. Teachers, through the HE input, have the opportunity to read, reflect, discuss and critique the social and political factors which influence the social context of the school. In this context they need to be deliberately challenged to refocus their practices through school-based research; to confront the existing assumptions underlying their actions, and; to be involved in converse debates about the issues researched. There is little opportunity for this to occur spontaneously in the school or organization, therefore a process of school and professional reconstruction is made possible using action research in a QD process supported by the intellectual challenges provided in higher education.

There are also responsibilities attached to the facilitation of reflective action related to scholarship in the process of participating in authentic collaboration with school-based educators. Ann Lieberman (1986) refers to the school–university partnership which she sees as a framework for restructuring schools and creating professional development schools as 'scholars of practice'. Her perception of the partnership indicates the value attached to cooperating in opening up issues from 'inside' schools for exploration 'outside'. The creation of schools for professional development is another exciting concept for research-based educational planning. Most significant, however, is the word 'scholar' linked with practice. The HE tutor in assessing the QD process, deliberates only on the professional's ability to perceive the situation in a balanced manner, to take action in a deliberated and weighted way, and to follow-up with a critical evaluation of the consequences. Can scholarship strengthen the process and make it more challenging and critical of existing practices?

What is Missing?

The QD process contains many worthy aims and espouses many highly laudable values related to participation and collaboration in the monitoring and evaluation of practice. It advocates an open management style, a responsive organizational structure and a whole school commitment. But it then includes in the methodology the stipulation that after the findings which result from monitoring and evaluation are used to inform school decision making and action, that there should be a documentation of the procedures and systems developed for this process to be subjected to internal review and to external moderation and validation (in preparation for Office for Standards in Education inspections?).

The suggestion is that the framework for action be examined by 'outsiders' and its validation depends on this approval. Another observation on the QD process is the absence of any reference to problems in the development of or failure to implement deliberate planned action. There is no envisaging of the research itself and the knowledge it bestows on the learner, as the 'action'. There is no reference to the eternal and irresolvable changing nature of school-based action research. The process appears to focus on the aim of a final product rather than investigation and reflecting on the dilemmas inherent in the school curriculum and the conflicting values of staff, teachers, pupils and parents.

As referred to earlier, the greatest problem I experienced in the QD TEC development was the eagerness that managers employed in target setting and identifying criteria. They hastened to do this in a conceptual and abstract manner rather than draw the criteria for improvement from their concrete experience of reality based on the reconnaissance or initial audit which included themselves and their interpretations of their roles. The 'action plan' became a creative and imaginative exercise which bore little relationship to the real situation. The plan *dominated* the possibility of a detailed analysis of the data and mitigated against the reflection which should be implicit in the deliberative progress to plan transformative practices. Without reflection and deliberation how does one know '*how to get there*'. Without further research strategies involving *ideas and intentions* rather than targets and plans, which of necessity involves negotiation and communication with others involved, e.g., pupils, parents, colleagues or external personnel, how does one judge the quality of the action planned?

The QD experience with managers indicated to me a modelling of the political practice of a strong Conservative state which upholds strong leadership, rapid executive action, low tolerance of opposition and little commitment to consultative procedures in the making of new policy. This view is supported by Grace, who believes that '. . . in a wider socio-political and ideological climate in which strong leadership is lauded, and in a new competitive culture in schooling in which rapid, entrepreneurial activity is believed to be essential for institutional survival, external factors work against the chances of collaborative school management and in favour of hierarchical chief executives' (1995). The QD process sets the scene for rapid action and in so doing may unwittingly create hierarchical management strategies for action through the school QD coordinator. There is *no reference* to the possibility of changing power relations or policies in education as a major or even minor consideration in QD endeavours. There is no reference to the importance of self-knowledge in educational development or the view that postmodern society must seek to learn to think about its own thinking. Kincheloe (1991) writes that, 'Researchers who do not understand themselves tend to misconstrue the pronouncements and feelings of others' (p. 45).

In QD there is no suggestion that there needs to be attention to the reconciliation of multiple interpretations of school situations that may be difficult to resolve. There may be implicit arbitrary rules at play in the situation that cannot be acknowledged without some resistance or confrontation. Is there a reliance upon certainty and prediction in organizing verifiable facts into a theory? How does one resolve contradictions? Where is the quest for insight into how assumptions have been constructed? Where is the attention to the process of question setting rather than to question answering? It is in the construction of questions that the teacher-researcher is creating a view of the situation that may be original and bold and subsequently may lead to innovative thinking and practice. Is there an opportunity to know what the researcher is thinking? What and where is the reflexive context of the self-evaluation?

The QD process emphasizes the world of objects — the management of change and people, rather than the intersubjective world of persons. There is no reference to the political and moral dimension of planned change. The guise of autonomy and responsibility in QD may perpetuate traditional practices by circumscribing autonomy, while at the same time relieving the actors/teachers of responsibility for planned actions. It is the hidden dispositions and dominant values within the school that must be made overt and this takes the QD coordinator beyond professionalism. Their professionalism is hidden within the culture of schools. How can we consider the educational context without acknowledging and understanding the culture that covertly supports it? The traditional culture of teaching predisposes the profession towards authoritarianism while, at the same time in adopting such an approach, the teacher is acting according to assumed acceptable professional values. Yet, in this action s/he is absolved from the responsibility of personally justifying the authoritarianism in their actions because they are legitimated through the sanctioning of professionalism in the work associated with, for example, managing QD in the school — if these issues are not questioned.

The professional practice of QD could then put the persons involved in *educational* practices, both teachers and students into powerless positions in the accepted face of traditional wisdom and practice. Without exposing the way in which such practices constrain the teacher's work we continue to disguise the lack of autonomy of both teacher and student in the mask of good and consensual practice which conceals the subtle distortion of the actions. To be liberated from such a distortion therefore requires a different interest from simply a practical 'improvement' or the attainment of a quality interest. It may require something closer to the emancipatory interest of Habermas (1984) or emancipatory praxis referred to by Grundy (1987) as radical professionalism or 'practique'. This allows for a deeper insight into everyday practice through a questioning and asking *why* things are the way they are and *why* they need to continue in this way? There is then the possibility of transforming 'ideologically frozen relations of dependence' (Habermas, 1984, p. 2). This necessitates an attitude or disposition which goes beyond accepted practices in traditional professionalism. It is the questioning attitude to managing QD in schools that will enable it to be emancipatory.

The process that has been omitted from QD is that of critical reflection. The reflection is related to a form of critical pedagogy which confronts the real problems in schools and are not pseudo-problems associated with the implementation of predetermined aims and objectives grounded in accepted professional wisdom. Critical reflection generates a transformation of awareness which is similar to what Friere called 'conscientization'. If the process is successfully developed in professional development contexts then there must be a deliberate attempt to raise awareness to a level that makes it possible to critique existing assumptions in underlying practice and offer some help in the means of changing the existing order of things. We may find in our judgments, reflection and evaluation that some particular action is essential to the improvement of the school situation. But is it the improvement of teaching or learning that we are aiming at? Is it the pupils or the teachers that the QD process is aiming to help? This becomes contradictory and unclear in the analysis of the QD procedures.

Ideological constraints will be realized in structures supporting educational communities when it is pointed out that the rules of the game *allow* this to happen. I believe it is the role of an 'outsider' or critical facilitator to enable school researchers to perceive covert educational structures. An ideological critique of school practice enables teachers and other educational professionals to recognize culturally derived meanings which are in principle open to change. It is only in the development of a critical educational community that an alternative discourse will emerge. It is also impossible to gauge the time needed to develop a consensus about meaning as a basis for collaborative action. Most programmes for teachers in HE based on either action research or QD are time constrained in some respects.

A critical interpretation of what is going on in schools controlled by practitioners expands the boundaries of professional practice to reveal *dysfunction* and creates solutions grounded in the *moral* basis of practice. Realizing the moral basis in practice allows teachers to become conscious of their values and how they affect practice. It allows them to formulate an argument for the support of better practice and to contradict through debate and discussion the institutional constraints which it is difficult for them to perceive without penetrative insight facilitated in the action research process. As teaching is a moral enterprise, critical appraisal of educational systems and institutional practices including the institutionalization of change itself is an essential part of reflection and enquiry. If we are following the precepts of Dewey then where is the ethical imperative to create a politicized profession capable of struggling for various forms of public life informed by a concern for equality and justice? And how do we encourage the existing tutors in the QD enterprise in higher education and in the local authority to take up the position of transformative intellectuals?

Critical pedagogy classifies schooling as a form of cultural politics, focusing an investigatory eye on the ways in which educational institutions perpetuate the interests of the dominant group while continuing to disenfranchise others on the basis of race, class and gender (Apple, 1992; Aronowitz and Giroux 1991). Critical educators advocate a language of critique and a language of possibility as the basis for reform, to begin and sustain a dialogue that interrogates and transforms the existing social order through democratic education.

It appears that a limitation of the QD implementation in education is its narrow emphasis on the technical, the operational and the measurable elements of educational practice which marginalizes the historical, political and socio-cultural aspects of the activity. The discourse and language of QD has confirmation an obscure view of comprehensive vision and free expression.

There is a danger that the QD slogan in practice will become reductionist and result in its over simplified application and use. It may be a case of reflection within of political forces beyond. It is Bernstein's (1990) belief that 'Education is a relay for power relations external to it'.

The age of market management is now the dominant discourse in the provision of all public services in England in the 1980s and 1990s. This new discourse has arisen from New Right ideological attacks on the weakness of social democratic principles which operated in Britain from the 1940s to 1970s when education was regarded as the actual means of effecting greater democracy for social justice. 'The language of choice and diversity in education has, in the 1980s and 1990s been appropriated to serve the interests of advancing a market culture in schooling' (Grace, 1995). Perhaps when the culture and practice of the new managerialism increases and becomes more public in schools in the 1990s, the principles of school management will be critical in the shaping of the school's ethos and its internal social relations. Shared decision making and the culture of professional collegiality could continue in a new form as advocated by Caldwell and Spinks (1988).

Self Knowledge and Growth

Dewey (1966) contended, with respect to moral growth, that 'it is impossible for the self to stand still; it is becoming, and becoming for the better or the worse. It is in the quality of becoming that virtue resides.' Becoming is based on self-knowledge and this is significant because, as Hansen (1993) points out, teachers' attitudes and bearing influence students' disposition towards others, towards themselves and towards the process of learning itself. A teacher may intend his/her actions to have a beneficial effect on pupils to help them become more confident and self-directed but is unaware of the unintended consequences of his/her efforts. Many of the moral meanings evident in classroom teaching are unintended because they result from the enactment of what the person is, which is more than their conscious intentions realize. So how can the teacher control, much less dictate, the impact they have in educational contexts without self-knowledge and understanding? Reflection on practice brings to the centre the exploration of the person's attitude and manner, brings to the teacher's attention the awareness of personal characteristics as an element in social interaction and the necessity to include it as an aspect of the improvement of conditions which make educational endeavours worthwhile. Dewey (1966) stated that it was more important to give practitioners the means by which they can become their own enquirers rather than telling teachers what to do. The narrative of self-identity is essential for professional growth in the rapidly changing circumstances of school development planning and the advancement of present government policies. The narrative can be best captured in a journal or diary which is accepted practice in most action research courses in HE, yet is not alluded to in the QD process as a significant aspect of the action research process. The action researcher or the QD coordinator in schools needs to learn how to, 'integrate information deriving from a diversity of mediated experiences with local involvements in such a way as to connect future planning with past experience in a reasonably coherent fashion' (Giddens, 1991).

Conclusion

As a result of the work pressures that teachers are experiencing from curricular and assessment reform in the UK currently, it may be that they are not positively motivated to actively participate in collaborative school developments even if they endorse its theoretical virtues. What the introduction of QD to schools has demonstrated is that school development is at a major cultural turning point. The established ways and practices in schools are changing and new patterns are emerging. The critical question is, what will shape these new patterns and what form will they take?

The democratization of education is a process based on a commitment to power sharing and the evaluation of school outcomes and achievements (see Simons, 1987). Any proposals for increased democratic involvement in education must involve finding the reflective space to investigate and evaluate the energetic and active practices in the everyday organizational strategies employed by those entrusted with the education of students of all ages.

I find a tension between the individual professional in a liberal educational context making decisions about changing practice as opposed to the whole school collaborative approach to supported self-evaluation. In creating a self-evaluative process within whole school development plans there is the possibility of realizing a democratic process in the educational world involving teachers and students. However, in a democracy, effective action must necessarily be based on voluntary contribution — taking place essentially through the art of persuasion. The implication for both the action research process and the QD process is that changing the educational contexts in which one works depends not so much on action but on the willingness to effect change that is authentically related to one's educational values. It also implies that learning to be persuasive is an art to be included in any methodological approach that supports deliberate change in schools using collaborative action. Finally, I believe that simply to follow the process of action research for school improvement is not enough in itself, without the expression and exposure of values towards others within the process. What is improvement of practice without acknowledged aims? What is collaboration in action research or QD processes, if it does not have a central ethic/principle acknowledged within it?

Action research is a learning process which begins with the individual professional researching educational issues in association with colleagues, students, parents and others. QD appears to be a means of delivering the School Development Plan in the most effective way. The QD coordinator, it seems from the evidence presented in the process, is not expected to have a personal view that is different from the school view. The QD coordinator is expected to attain consensual agreement about the overall school QD aims. One cannot be in opposition to the school's aims and at the same time deliver the mission, build the organization, and support the implementation of strategies if they are inconsistent with one's own values and views. This observation has implications for the initial selection of the QD coordinator and the smooth running of the QD change and development process in the school. The constructive process aimed at through QD may be viewed as otherwise by participants if the coordinator is selected for his/her transformative and high risk-taking qualities! How does one implant a reflexively ordered narrative of self-identity into the educational process of professionals when ends are the ultimate aim? How does one become constructively involved in the politics of the school if the politics are inconsistent with one's personal values and the consensual practice contradicts one's own?

The action research process for me, in comparison to the QD process outlined earlier, retains its integrity because it is flexible, challenging, allows openness and honesty in encouraging the exposure of the real life in schools to possible transformation, and because its agenda for reconstructing educational contexts is totally open. It is the difference between working within or beyond the constraints of education systems. Action research which is educational encourages the researcher to go *beyond* the constraints imposed by schools and to act for the reconstruction of educational systems. The QD process directs its proponents to work *within* the constraints to improve the existing system in whatever manner is thought to be most effective. In going beyond the constraints, the self-growth of professionals in education is vital to the development of institutions and communities, and integral to the process of altering educational consciousness. Without it any educational progress, even that based on practitioner research, will alienate its participants.

References

APPLE, M. (1982) *Education and Power*, Boston, Routledge and Kegan Paul.
APPLE, M. (1992) *The Politics of Official Knowledge*, New York, Routledge.

ARONOWITZ, S. and GIROUX, H. (1991) *Post-modern Education: Politics, Culture and Social Criticism*, Minneapolis, University of Minnesota Press.

BIRMINGHAM EDUCATION DEPARTMENT (1993) *Quality Development Resource Pack*, Birmingham, Birmingham City Council.

CALDWELL, B. and SPINKS, J. (1988) *The Self-Managing School*, London, Falmer Press.

CROSBY, P. (1970) *Quality is Free*, New York, McGraw Hill.

CROSBY, P. (1984) *Quality without Tears*, New York, McGraw Hill.

DEMING, W. (1982) *Quality, Productivity and Competitive Position*, Cambridge Mass, MIT Centre for Advanced Engineering.

DEMING, W. (1986) *Out of Crisis*, Cambridge Mass, MIT Centre for Advanced Engineering Study.

DEWEY, J. (1966) *Democracy and Education*, New York, Macmillan/Free Press.

ELLIOTT, J. (1991) *Action Research for Educational Change*, Milton Keynes, Open University Press.

FULLAN, M. (1992) *The New Meaning of Educational Change*, London, Cassell.

GRACE, G. (1995) *School Leadership: Beyond Education Management*, London, Falmer Press.

GIDDENS, A. (1991) *Modernity and Self-Identity*, Cambridge, Polity Press.

GOFFMAN, E. (1959) *The Presentation of Self*, New York, Doubleday Anchor Books.

GRUNDY, S. (1987) *Curriculum Product or Praxis*, London, Falmer Press.

HANSEN, D. (1993) *From Role to Person, American Education Research Journal*, **30** (4), pp. 651–75.

HABERMAS, J. (1984) *Theory of Communicative Action*, Boston, Beacon Press.

HOLLY, P. (1986) 'Developing a professional evaluation', *Cambridge Journal of Education*, **16** (2).

HOPKINS, D. and AINSCOW, M. (1993) *School Improvement in an ERA of Change*, London, Cassell.

ISHIKAWA, K. (1985) *Guide to Quality Control*, Tokyo, Asian Productivity Organisation.

JURAN, J. (1979) *Quality Control Handbook*, New York, McGraw Hill.

JURAN, J. (1980) *Quality Planning and Analysis*, New York, McGraw Hill.

KINCHELOE, J. (1991) *Teachers as Researchers: Qualitative Enquiry as a Path to Empowerment*, London, Falmer Press.

LIEBERMAN, A. (1986) 'Collaborative research: Working with, not working on', *Educational Leadership*, **43** (5), pp. 28–32.

OAKLAND, J. (1986) *Statistical Process Control*, London, Heinemann.

PETERS, T. (1982) *In Search of Excellence*, New York, Harper and Row.

REYNOLDS, D. (1992) *School Effectiveness: Research, Policy and Practice*, London, Cassell.

SIMONS, H. (1987) *Getting to Know Schools in a Democracy: The Politics and Process of Evaluation*, London, Falmer Press.

SHULMAN, L. (1987) 'Knowledge and Teaching', *Harvard Educational Review*, **57** (10), pp. 1–22.

TAGUCHI, G. (1979) *Introduction to Off-line Quality Control, Central Japan Quality Control Asscnn*, Magaya, Japan,

TECHNICAL, VOCATIONAL AND EDUCATIONAL CENTRE, BIRMINGHAM EDUCATION DEPARTMENT (1990) *Developing Quality Through TVE: Staff Development Resource Pack*, Harborne, Birmingham, TVE Centre.

6   Seeing-off Cuts: Researching a Professional and Parents' Campaign to Save Inclusive Education in a London Borough

Alice Paige-Smith

ABSTRACT *In this paper I firstly present an action research case study of a parent and professional campaign. Secondly, I critically discuss the dilemmas of conducting action research with parents in my role as a researcher, parent and advocate.*

Introduction

This paper is about a campaign by parents and professionals to save the support service for statemented pupils integrated into mainstream schools in a London education authority. I consider the development of the campaign and the process of researching parents' experiences in this campaign.

A campaign by parents and professionals was launched in March 1991 to save all thirty-eight special needs assistants and six support teachers. The council's Education Committee proposed to axe these posts as part of £5.5 million savings in its 1991 education budget. One hundred and four pupils who experienced difficulties in learning or had disabilities were to be affected by this decision. Support in the form of a welfare assistant and/or a support teacher was to be removed from these pupils in mainstream schools. This extra support was provided to these pupils by the learning difficulties service and the emotional and behavioural difficulties service. In 1991 pressures from central government rate-capping policies resulted in the local authority having to make cuts in education services. At the same time the local authority was also obliged to delegate budgets to schools under the 1988 Education Act, and the Department of Education and Science regulations issued in December 1990 on the local management of schools (Lee, 1992). A firm of financial consultants facilitated the local education authority's decision to reduce the learning support service.

Ann and Tom were two members of the core group of parents. They joined the campaign at the first Open Meeting. Their son was in the same class as my daughter. Ann turned to me for advice and support when the campaign started. Once I had introduced myself to the professionals involved in the campaign I was invited to all the meetings which they organized and attended. I was included into all the strategic planning meetings. I provided legal advice, and they asked me to write letters to support the parents whose voice they required in order to retain the service. I tape-recorded most of the meetings. I also kept a 'diary' and wrote up the sequence of events as they happened. All the campaign documentation written by parents and professionals was collected and retained. Newspaper articles were also collected, parents were given a questionnaire to fill out and some parents were interviewed.

Firstly, I shall consider the background to the campaign and outline the development of the support service within the authority. This service was set up following a review of special educational needs provision within the authority which drew on parents' perceptions. The consequences of implementing the cuts to the support service are examined.

Secondly, I shall consider how the campaign was initiated by the learning support service who enlisted the help of a disability organization, the Disability Consortium, within the authority, an organization for disabled people and children, then informed parents about the cuts to services. The professionals from the Disability Consortium and the learning support service organized the deputations and lobbying by parents to committees and an MP in the local authority. However, the demarcation between who was a professional and who was a parent blurred during the deputations. Some of the professionals were also parents of children to be affected by the cuts. The professionals knew whose opinion to influence and when. They were also in touch with local parent associations and other professionals who were concerned about the cutbacks in education services. The campaign to save the support service was also a part of a larger campaign within the authority. The campaign culminated with the Education Committee meeting when councillors set the budget for the following year.

Thirdly, I discuss the issues arising from the case study of this parent professional campaign. The issue of whether the parents and professionals have the same aims in the campaign is considered alongside the overlap between. The role of the professionals is also looked at critically. The successes and failures of the campaign are considered. Finally, I consider the conflict of identity which occurred in my role of conducting 'praxis orientated research' with parents and professionals.

Background

Setting Up a Support Service

The 1981 Education Act placed a duty on local authorities to review their special education provision. Section 2 of the 1981 Education Act states that:

It shall be the duty of every local education authority to keep under review the arrangements made by them for special educational provision. (Department of Education and Science, 1981, p. 2)

Despite this duty placed on local authorities, not all local authorities looked at their special education provision. In 1987 information was collected from 105 local authorities about their reviews. Of the 74 replies 46 per cent had completed reviews of their special educational provision (Allen, 1987). The importance of conducting a review of provision and the formation of policy in local authorities on integration was emphasized in the 1987 Education, Science and Arts Committee Report 'Special Education Needs. Implementation of the Education Act 1981':

Our experience is that it is important for a LEA to have a clear statement of its policy on integration, developed in consultation with parents and professionals, as part of its overall policy on special education. (Education, Science and Arts Committee, 1987, p. 12)

In 1987 the report entitled 'A Red Bus Next Year? Consumer Perceptions of Special Education in the London Borough of Haringey' was produced by the council's education service. This report contributed to the review of provision recommended by the 1981 Education Act. Concern was expressed in the report over the exact form and process of the integration of pupils:

The 1981 Education Act has brought about nationwide changes in special education policies and provision . . . However, despite clear improvements, these developments have also led to a number of anxieties: for example, concern over the exact form and process of integration of special needs children into mainstream facilities. (Berridge and Russell, 1987, p. 1)

Consequently statistical information on the number of pupils requiring provision as well as the parents' and pupils' perceptions of their situation and services was considered desirable. The report consisted of interviews with 100 parents. A 30 per cent response rate was received from the 861 questionnaires sent to all of the parents of pupils identified as requiring special provision. This report had similar findings to the Inner London Education Authority's Fish Report (1985) and the Education Science and Arts Committee (1987). The report found that parents were experiencing difficulties in the assessment process and its consequences. Parents reported that they had received insufficient information on what special help might be available. One parent complained of 'inaccurate assessments', others had 'fought long battles to get special help' and once additional help was identified parents reported that it was 'slow in coming' (Berridge and Russell, 1987). Support was not provided for parents from any voluntary organization or advocacy group for parents, and parents had not had contact with a 'named person' as recommended in the Warnock Report (DES, 1978). Despite the overrepresentation of 'minority ethnic communities' in 'MLD' (moderate learning difficulties) and 'EBD' (emotional and behavioural difficulties) and 'SLD' (severe learning difficulties) provision the report notes that 'the large majority of parents seemed satisfied with their child's school and the arrangements made for his or her special needs.' However, parents reported concerns over protracted assessments, physical access to schools for disabled pupils, the quality of education, general educational standards, and lack of training in social skills. Whilst some parents expressed a preference for special schools, others welcomed the philosophy of integration, although, on the whole, parents referred to their own particular schools and their own experiences.

Having made the effort to consult with parents, pupils and eight working groups of professionals, the authority wrote its review of special educational provision in 1987. A number of principles including the maximum possible integration alongside the encouragement of parental involvement underpinned the recommendations in the review. Provision for children with 'complex and severe needs' formed four services grouped around: learning difficulties; emotional and behavioural difficulties; sensory and language difficulties, and; physical and medical difficulties. Support services for pupils in mainstream schools who have a statement outlining their needs and resources were set up by 1991 in the form of a learning difficulties service and the emotional and behavioural difficulties service. The teachers and special needs assistants were managed and coordinated in the same premises by the two separate heads of the learning difficulties and the emotional and behavioural difficulties support services.

Trimming the Budget

In February 1991 the primary aim of the council was to reduce spending. Their budget had been reduced due to central government policies of poll-tax capping. In 1991 some councils were under attack by the government over the way they spent their money. A letter written by Norman Tebbit MP indicates the strength of feeling towards left-wing councils by the government. He was responding to a letter which expressed concern about the cuts in the authority sent to him by the clerk to the school governors.

> As you know, your Education Authority spends more per pupil and has a worse educational record than almost any other in the country. If more money is required (which frankly you have not established) there is plenty of money being wasted by your Council. I suggest you resolve your difficulties by performing as well as, say, the average of the third least successful quartile of education authorities and if that needs money, by persuading your Council to spend more on education and less on loony left idiocies.

Each council service comes under the control of a committee, made up of councillors, and co-opted members of the community. The Education Committee had conflicting aims in 1991. On the one hand they had to run an education service on a reduced budget for the following financial year, and on the other hand they were obliged to provide a service to the 'consumers' of that service according to the existing educational policies.

The total savings across the council were proposed to be £22 million. The Education Committee proposed to make savings of £5.5 million in its 1991 education budget. Out of this sum the Education Committee anticipated that savings could be made in the authority's educational provision for children with 'special educational needs'. The authority had been advised to take this decision by a private firm of consultants called the P.A. Consulting Group, based in Hertfordshire. These consultants, hired by the education service, facilitated decisions concerning budget allocation for the delivery of statutory local education authority services. The head of the learning support service recognized that the Education Committee's decision was also influenced by the Coopers and Lybrand report to the Department of Education and Science on the Local Management of Schools (1988). This report recognized the proposition in the Education Reform Bill (1988) that there should be 'greater delegation to schools of responsibility for financial and other aspects of management' (Coopers and Lybrand, 1988, p. 5). Further regulations were issued by the Department of Education and Science in December 1990 to delegate 90 per cent of their budgets to schools under the arrangements for Local Management of Schools. Tim Lee (1992) notes that these regulations were 'designed to force LEA's further into "shifting resources from administrative overheads and other central services to school budgets"' (Lee, 1992, p. 164). The head of the learning support service was also aware that school governors were telling the LEA to devolve the budgets for support services directly to the schools. However, the Education Committee failed to recognize that they would be acting illegally if support was removed for pupils with statements of their 'special educational needs'.

The Parent and Professional Campaign

Chronology of the Campaign

In February 1991 the council's Education Committee prepared to remove the financing of all thirty-eight welfare assistants and six of the support teachers who supported 104 pupils integrated into mainstream schools. The Disability Consortium — a disability advice centre funded by the council — were alerted to this situation by their co-opted member on the Education Committee. She informed the learning support service and together they decided to inform all the parents of children who receive support in the authority about the threat to the future of integration. A letter outlining the cuts was written by the

Table 6.1: Current costs vs potential cost of alternative provision

First child	Second child	Third child
Cost of current provision	Cost of current provision	Cost of current provision
Special Needs Assistant — £2,000. Teacher — £750	— part-time assistant — £1,000	Special needs assistant — £4,000. Teacher — £750
Total £2,750	Total £1,000	Total £4,750
Cost of day special school — £11,000 including travel	Cost of day special school £11,000 including travel	Cost of day special school — £11,000 plus travel
Residential placement — £30,000 including travel		Residential placement — £30,000 plus travel

Disability Consortium and posted by the support services who had the names and addresses of the parents. This letter invited parents to an Open Meeting (6th March 1991). At this meeting a core group of parents were asked to volunteer to represent the parents of pupils who were to lose their support. This core group of parents alongside representatives from the Disability Consortium and the learning support service attended a deputation made to the council's Labour Group (7th March 1991). This deputation then went to their local MP (12th March 1991). The final deputation was made to the Education Committee (21st March 1991). After this final deputation an Open Meeting (17th April) was organized by parents, the Disability Consortium and the learning support service.

Examining the Figures

The decision of the Education Committee to reduce support services did not make financial sense. It also contradicted the authority's written policy of integration, it was unlawful and caused untold stress and upheaval to the families, the children themselves, their special needs assistants, the schools they worked in and the support services.

By removing all thirty-eight special needs assistants and six support teachers the council would be saving £480,000. The consequence of making cuts in this area would result in the children being sent to special schools or units. The paradox is that this action would be more expensive. The actual figures for five children were costed out during the campaign (with the consent of the child's parents) to be on average four times as much for each child (see Table 6.1). The following costs for 3 children show the expenditure on their current provision versus the potential costs to be incurred if they were transferred to a special school or residential placement.

The task of the campaign was to make councillors aware of the financial contradictions: that it was going to cost more in the long term if the funding for the special needs assistants and support teachers was removed.

Enforcing *Legal* Duties

Some of the children who were to lose their special needs assistant have this provision outlined as essential for their education on their statement of special educational needs. This document, drawn up by a team of professionals, outlines the 'special education provision' required by a child in Section 2. For example, Section 2 of one child's statement was set out as follows:

SECTION 2 SPECIAL EDUCATIONAL PROVISION
A mainstream setting which is able to offer normal models of language and behaviour.

His progress should be closely monitored. The focus of support should be an individualized programme devised by the Class Teacher in collaboration with the Support professionals. In order for such a programme to be implemented, the teacher would need the support of an additional welfare assistant from the Learning Support Service.

The Disability Consortium and the learning support service recognized that if the education authority failed to provide what Section 2 specifies then the parents have the legal right to demand this provision through the courts. The campaign culminated by threatening to take the education authority to court if they attempted to remove provision that children are entitled to by law. However, only those children who have statements which clearly outline this provision were legally protected. As this local authority has a policy of 'low statementing' and many children may not have had statements to protect their provision, the goodwill of the authority is assumed, in so far as they are expected to provide the *children* with the support they need without it being written in the form of a statement. So those children without a statement were in a more vulnerable situation because they did not have a legally binding document.

Unfortunately the Education Committee had to be made aware of the financial 'cheapness' of integration. They also had to be made aware that it is illegal to remove children's support if it is outlined on their statement. It was planned to inform councillors that those children without statements would all be requesting them from the authority. This action would cause administrative upheaval and disruption to the education offices.

The Open Meeting: Mobilizing Parents

On the 5th March the Disability Consortium sent a letter to the parents of pupils who received support in the authority. This letter informed parents about the cuts in provision and invited them to an Open Meeting on the 6th March. The Open Meeting had a better response than was expected, the small circle of chairs set out in the centre of the room grew to a larger circle, and then the circle disappeared. Rows of parents and professionals filled the room, which became full to capacity. The support teachers and special needs assistants had received letters offering them the option of voluntary redundancy. They knew that their posts were under threat.

The meeting was coordinated by professionals from the learning support service, the inspector for special educational needs, the Disability Consortium and included two parents, one who was a special needs assistant, the other the parent representative to the learning support service. A headteacher, the inspector for special needs, and a special needs assistant all spoke about how the removal of the special needs assistants would affect integration in the authority. Moira Morgan, the co-opted member for the Education Committee from the Disability Consortium stated that:

> Haringey has no equal opportunities policy for children with disabilities, there is no equality of opportunity for adults with disabilities, we have to make sure they are catered for through primary education, through secondary education into further education. We have to make sure that councillors ensure that there is a service there throughout their school lives. A commitment to integration forms a part of the education policy of Haringey, but this is not fulfilled in reality. Haringey's Policy for Special Needs written in 1987 states that: 'As far as is possible, children and young people should be members of mainstream schools and colleges; segregated provision should increasingly give way to supported integrated provision. Provision for special educational needs should be seen as being part of one comprehensive system of education sharing a common set of aims for all pupils, working within an agreed but flexible curriculum framework.'

A parent in the audience spoke out about the effect the cuts would have on her child with Down's Syndrome who has been in her local school for years. She posed a question to the councillor:

> Who is going to tell my child that she could not go to her school anymore, what do I tell her in three weeks when I can't walk through the door with her?

This parent had been informed by her child's headteacher that as of 31st March her child would not be able to attend her school due to health and safety reasons if she did not have her special needs assistant. As this parent was also a special needs assistant whose job was under threat and had been involved in setting up the meeting, she was making the other parents aware of the worst scenario that could happen in their schools. A councillor had been invited as it was anticipated that he would support the campaign. It was hoped that he would listen to the consequences of the cuts his Labour Group and the

Education Committee had proposed. He responded by suggesting that the cuts were a direct result of central government policy, but the parents at this meeting were laying the blame on the council's decision to axe this service.

An advice worker from the Advisory Centre for Education, invited by the Disability Consortium, informed the meeting that the authority's proposals were unlawful. She argued that the authority would be in breach of their statutory duty if those children who had statements which outlined in Sections 2 and 3 that they should be educated in a mainstream school with support were to have that support removed under the 1981 Education Act.

A number of strategies were proposed at the end of the meeting. Firstly, to gather together the parents whose children had statements which outlined the above provision and to provide case studies of these children to the councillors to show them that if they had their support removed they would have to be segregated into special schools which would be more costly to the education authority. Secondly, it was proposed by the Disability Consortium and the learning support service that parents could threaten to request statements for their children who received support but did not have it written into a statement. Whilst this action was endorsed by the education officers it was hoped that the threat of parents demanding statements would ensure that the council would make a u-turn.

Richard, a parent involved in setting up the campaign, asked for volunteers from the audience to stay if they wanted to become involved in the campaign. Over the next two weeks these parents made a number of deputations to the Labour Group, school governors, Bernie Grant (the local MP) and the Education Committee. The deputations were conducted alongside other parents, teachers, and parent-teacher associations as the cuts were to effect the music service, and school meals.

Discussion of the Issues Arising from the Campaign

The case study of the campaign raises a number of issues about the education of pupils who experience difficulties in learning or have disabilities. The case study illustrates how difficult it is for councils to resource a policy commitment to integration. As the Education, Science and Arts Committee found, the implementation of the 1981 Education Act depends upon the policy commitment of the local authority.

> It seems clear to us that a successful implementation of the 1981 Act is very much dependent on the development by an LEA of a clear and coherent policy, arrived at in a way that enables it to command the support of those — parents, teachers and voluntary organizations — who are most affected by it. (Education, Science and Arts Committee, 1987, p. 18)

The difficulty in resourcing policy commitments is identified by the Committee. The pressures on local authorities to resource both special schools and support in mainstream are recognized by the Committee:

> The Committee is in little doubt that a lack of resources has severely hampered the successful implementation of the 1981 Act. Extra resources have been needed to fund the administrative costs of LEA's of the assessment and making of a statement as well as the major costs, which are difficult to quantify, of arranging more properly funded placements in primary and secondary schools. The latter costs arise not only from teacher or ancillary staff costs but also from equipment and physical adaptation of premises. Resources will be under particular pressure where Local Education Authorities are providing alternative forms of provision, in both special schools and in primary and secondary schools, for similar forms of special educational need. The Committee concludes that the lack of specific resources has restricted implementation of the 1981 Act. A commitment of extra resources is needed if significant further progress is to be made. (Education Science and Arts Committee, 1987, p. 13)

The case study illustrates how the 1981 Education Act has been difficult to implement at a local authority level. Whilst the authority in the case study managed to write a policy on special education it was unable to uphold its commitment to integration. This was due to financial pressures from central government; the education authority had to reduce their education budget for 1991. The case study illustrates how the integrated education service was considered to be 'alternative' and 'dispensable' by the Education Committee, in contrast to the special schools which were not affected by the cuts in education. The Education Committee had been wrongly advised by consultants hired by the local education authority to make the 'marketed' decision to reduce the learning support service. When these consultants helped to facilitate the budget allocation they failed to recognize that pupils with statements have their provision legally protected.

The campaign arose out of a specific issue: the axing of all educational support for children integrated into mainstream schools. The lack of commitment to integration is not isolated to this specific campaign. The underfunding of support services, inadequate training for mainstream teachers, and lack of schools adapted for children with physical disabilities has been recognized by the Audit Commission (1992) and the Education, Science and Arts Committee (1987). Pupils in special and mainstream schools who have statements of their special educational needs were treated differently by the council. Those pupils who have statements but attended mainstream schools were not going to have their provision maintained by the council. Pupils statements in the case study had to be used by parents to protect their children's right to provision.

Orchestrating the Campaign

The aims of the campaign orchestrated by the professionals was to inform the parents, the 'consumers' of the service, in order to save their jobs. The parents then had to inform the councillors of the importance of the service and their legal obligation to fund it. The dilemma faced by the support service was that they were unable to represent their views directly to the parents about the decision to make cutbacks. Consequently the support service used the disability rights organization in the authority.

The campaign brought together different groups with differing interests. These different groups were united by the same aim; to stop the cuts in the support service in order to save the integrated placements of pupils. The parents wanted their children to remain in mainstream schools with support. The Disability Consortium supported this aim on a disability rights level. It had been this organization which launched the campaign by alerting the support services and parents. The professionals from the learning support service were primarily concerned with the retention of their posts and the service which they worked in or managed. The mother of a child who received support and was also a professional in the learning support service was involved in the campaign as a parent and a professional.

The case study illustrates the importance of having locally based committees which allow the consumers and providers of a service to represent their views on decisions made about education services. Through the deputations to the Education Committee, parents, professionals and disability rights workers managed to secure the support service.

Manipulation of the Parents by the Professionals

The professionals who participated in the campaign used their power, knowledge and authority to empower parents to speak on behalf of the learning support service. The professionals manipulated the parents into thinking that their own child's provision was under threat and could be saved by direct action. However it was the support service that had been saved, individual pupil's support in mainstream was not guaranteed after the campaign unlike the assistant and most of the teaching posts.

Parents attended the first Open Meeting organized by the professionals because they were concerned about the future of their children's integration. The letter sent to the parents on the Disability Consortium's headed paper, but written by the professionals urged the parents to attend the Open Meeting because the cuts threatened the future of their children's integration. At this Open Meeting parents were led to believe by the professionals that their children could be excluded from schools due to health and safety reasons if they did not have their support in mainstream. A headteacher and a special needs assistant — also a parent — spoke about the imminence of the exclusion of their children. Whilst the exclusion of pupils could have been a consequence of the cuts it is more than likely that the headteacher might have found a solution to keeping that child in school. However, it was within the power of that headteacher to refuse to admit a child who did not have the support required. It was also within the power of the parent to say 'What can I do?' when her child was threatened with exclusion. A scenario was made up between the headteacher and the special needs assistant/parent to shock parents into the reality of their situation. If the cuts were imposed the choices for parents were to either, 'keep their child at home or send them to a special school'.

The Successes and Failures of the Campaign

The campaign was successful because the budget for the support services was retained. The campaign to save the support services raised the profile of the importance of integration in the authority. The legal, moral and education policy issues of integration were publicized to parents, teachers, education officers and councillors. Since the launch of this campaign the support services have been perceived by the councillors to be a part of the core budget of education services in the authority.

The parents who were empowered to join the campaign were led to believe that they were protecting their child's right to integration. This core group of parents put a lot of effort into the campaign. Despite this effort one mother was disappointed at the inability of the managers of the support services to be her ally when she needed them. The segregation of her child into special school was a direct consequence of the support services being unable to provide support for her child in a local mainstream school due to the lack of funding for the learning support service. The campaign had successfully maintained a status quo in the support services and the practice of integration in the authority in the interests of the professionals. A consequence of this was that the provision of support, as allocated by the support services, was maintained at a level decided as appropriate by the support service, educational psychologists and education officers, not parents.

Conducting Action Research in the Campaign

A Conflict of Identity

During this piece of action research I had to consider how I should get involved and immersed in a research setting and what role I should present in order to become an 'insider'. I found myself asking questions: Was I doing research with parents as a professional teacher exploring educational issues, as an advocate facilitating parental involvement and rights, or as a researcher intent on finding out everything there is to know from the perspective of parents in the research? These roles often merged in the research setting.

I had been asked by a fellow parent about parents' rights to integration at my daughter's school. This parent, Ann, knew I was doing research in this area as we had talked about it together. I thought about how as 'a parent' I could advise her, yet feared getting labelled as a trouble-maker parent in the school if I became involved. As 'a researcher' I thought about what information I could gather from them. I wished at times that I could have whipped out my tape-recorder or that I had hidden it under my coat.

On the first morning when Ann and another mother, Julie, found out about the proposed cuts they strode into the headteacher's office and demanded to know what was going on. I stood by and supported their demands, explained the illegalities to the headteacher and then rushed off to photocopy the documentation to use in my research and to write up what had happened. I felt the conflict of roles particularly acutely when I was involved in a situation as a participant, but I had to document what was happening at the same time. I wanted to keep records of the incidences, conversations held at meetings, and informal chats between parents and professionals. However this proved difficult as I was contributing to these situations, rather than just observing and documenting them. A diary of events was used to collate the information in this case study. The relevant documentation was retained and daily events were recorded. In my role as a participant I felt restricted by my role as a researcher, I was aware that others would consider that my agenda was to extrapolate information for my own uses, rather than for the benefit of the campaign. I tried to counterbalance the possibility of hostile feelings towards my involvement as a researcher by contributing to the campaign. I did this by collaborating with the requests of the professionals who wanted me to ensure that the parent I knew participated in the lobbying of politicians.

My role as a parent, professional, and researcher was hard to combine and I felt like a tabloid reporter onto a good story. As the research developed so did this side to myself as a researcher. In my role I have been privileged to use my time as a researcher to be an observer, an active participant, and a recorder of events. The benefits of being a 'praxis orientated researcher' are outlined by Lather (1986). In her article she suggests that the researcher, interacting in the research process, can 'increase awareness of the contradictions hidden or distorted by everyday understandings . . . it directs attention to the possibilities for social transformation' (Lather, 1986).

Research has been carried out with the aim of assisting in the participants' understanding of their situations (Oakley, 1981; Elliott, 1991). An aim of action research has been defined by Elliott (1991) to 'improve practice rather than produce knowledge. The production and utilisation of knowledge is subordinate to, and conditioned by this fundamental aim' (p. 48). Rather than having carried out research on parents, to predict and control their behaviour, research has been done with them as I have considered them to be 'active subjects empowered to understand and change their situations' (Lather, 1986).

References

ALLEN, L. (1987) *Duty to Review*, London, Centre for Studies on Integration in Education.

AUDIT COMMISSION (1992) *Getting in on the Act*, London, HMSO.

BERRIDGE, D. and RUSSELL, R. (1987) *A Red Bus Next Year? Consumer Perceptions of Special Education in the London Borough of Haringey*, London, National Children's Bureau.

COOPERS and LYBRAND (1988) *Local Management of Schools*, London, HMSO.

DEPARTMENT OF EDUCATION AND SCIENCE (1978) *The Warnock Report*, London, HMSO.

DEPARTMENT OF EDUCATION AND SCIENCE (1981) *The 1981 Education Act*, London, HMSO.

EDUCATION, SCIENCE AND ARTS COMMITTEE (1987) *Special Education Needs: Implementation of the Education Act 1981*, London, HMSO.

ELLIOTT, J. (1991) *Action Research for Educational Change*, Milton Keynes, Open University Press.

INNER LONDON EDUCATION AUTHORITY (1985) *Education for All*, London, Inner London Education Authority.

LATHER, P. (1986) 'Research as praxis', *Harvard Education Review*, **56** (3), pp. 257–77.

LEE, T. (1992) 'Local management of schools and special education', in BOOTH, T. *et al.* (eds) *Policies for Diversity in Education*, London, Routledge.

OAKLEY, A. (1981) 'Interviewing women: A contradiction in terms', in ROBERTS, H. (ed.) *Doing Feminist Research*, London, Routledge and Kegan Paul.

7   Redefining, Reconstructing and Reflecting on the Group Process

Jane Richards

Teachers are always searching for the perfect solution to improve the learning process. In my search for better techniques, I've observed that cooperative learning is an appropriate procedure; theoretically, it engages all the participants. The process involves students who are working in small groups on a problem which requires the application of thinking skills. However, cooperative learning does not work for all participants in every situation.

Redefining

I teach in the English department at Cortland College, which is part of the State University of New York. Over the period of two spring semesters in 1993 and 1994, I have applied cooperative learning to feedback groups in the second segment of a freshman writing course. During the first segment, Composition 100, the students engaged in personal writing in addition to learning and practicing the conventions of quoting, paraphrasing and summarizing. A comparison and contrast essay was also assigned. In Composition 101, students wrote short and long argument papers in which they synthesized material from sources. As the students wrote, they followed the process of prewriting, drafting, conferring, sharing, revising, proofreading, editing, reflecting, and publishing. In addition the students read articles from a text based on multicultural issues.

The students who attend Cortland College originate from a variety of small towns and cities in New York State. In their high school setting, about 80 to 90 per cent of them have worked in groups; however, many of them had not participated in feedback groups to discuss drafts of other students' papers. I decided to conduct classroom research because I had used feedback groups in high school and Composition 100 courses with limited success. I felt encouraged when I read an article entitled 'What is Collaborative Learning?' by Barbara Leigh Smith and Jean T. MacGregor. They describe writing groups in the following way:

> This is a challenging process, one that requires students to read and listen to follow students' writing with insights, and to make useful suggestions for improvement. (1992, p. 16)

In addition to cooperating in groups, students must learn to respond to other writers' opinions and thinking patterns. I knew what I wished to accomplish, but I wasn't sure how to proceed.

Between semesters in 1993, I formulated the following questions to help facilitate the process of group activities and behavior:

- What are the benefits of feedback groups?
- What are the drawbacks?
- What is the teacher's role?
- How can the teacher and students work together to improve the process?

Teacher researchers and other educators have advocated cooperative learning for helping students become more independent learners. One of the foremost advocates of this technique is Dr William Glasser. A psychiatrist, who advocates reality therapy, Dr Glasser has applied this theory to education. He says, 'The idea of having students function as a group to produce some result has been carefully studied, and it works' (Gough, 1987, p. 659). I agreed with Glasser's general pronouncement that groups work, but from experience I knew that certain group situations were more productive than others. The following is a scenario which could have taken place in my classroom:

'All right, divide up into groups of four. Follow the directions written on the chalkboard.'

No one moved. Hesitantly, Nancy raised her hand. 'Do we have to work in groups?' Ray said, 'We did this in high school.' Tom adds, 'Why don't we save time and you tell us what's wrong with our papers, or,' he smirked, 'we can just go around the circle discussing what we did last weekend like we always do.'

A somewhat exaggerated account, but I had encountered resistance to feedback groups each semester. No matter what tactic I tried, I could never sell the students the idea that they are helping themselves as they gain more knowledge about how others write. I've used the reassuring philosophic stance of 'I'm here to help you — please try it my way' or 'Try it; I'm sure it'll be helpful.' However I always knew that the process could have been more successful.

In the past, I'd thrown students together in a group and watched them mix and match. I'd never discussed guidelines with them or modeled group behavior. I'd just assumed they knew my expectations, and they would immediately glean the benefits through sharing suggestions on their papers. My way didn't work for many reasons. Some of the students had participated in groups, while others had not. In addition, the first semester of college is an exciting time and socializing was infinitely more important than discussing possible improvements on papers. Also, I'd never described my expectations directly and never modeled positive group reinforcement. I'd provided the guidelines for group behavior without consulting the students.

Therefore in the spring of 1993, I concentrated on a classroom research project in which I constantly asked students for their opinions about the group process. On the first day of class, I followed the routine of introducing myself, the course, the text, and the assignments. Then I described the classroom research project in the following manner:

> This semester, I'm interested in conducting classroom research, but it won't be confined to research alone. That's why it's called classroom research. In order to find some solutions to the problem, I'm going to be asking you frequent questions. We'll be discussing your reactions, and I'll also ask you to fill out inventories or questionnaires and write JAWS (journals about writing). Now we don't know each other very well yet, but I hope you'll be honest with me when I ask you what's going on.

As I looked around the room, twenty-five students were listening intently, but when I asked if they had questions, no one volunteered. I continued to provide an overview of the problem.

> I've always been puzzled about working in groups. I'd like to know how people feel about working together. I want to know how we can improve the feedback process.

After we'd discussed the group process, we continued to work together in the large group for the remainder of the session.

To help me understand and adjust the process, I tracked the groups through my reflection journal during the spring semester of 1993. In one entry, I commented,

> All group[s] worked well (diagram to show organization) near window, sort of splintered, others formed a circle. They went right around the group and enlisted ideas from everyone.

One of my first notations had included the structure instead of the content of the group as I wondered whether the size of the groups made an impact on the way the group functioned. I discovered that the major disadvantage of the large group is time. All students had the opportunity to speak, but their activities were limited; whereas in small groups, people could speak more often and actually discuss issues.

Students also voiced their opinions about the group process. Roy said he was very uncomfortable in the group setting, while Jon commented on the personal benefits of the process:

> After sitting in the groups and evaluating other people's essays I picked up many addition ideas that will help me to write a better 3rd draft.

Both Roy and Jon belonged to the same group in which the members didn't always adhere to their tasks. Their responses seemed to indicate that the group process was working for one, but not the other. At that point, I interviewed my colleague, Vicki Boynton, who had worked with groups for many years.

In our interview, Vicki indicated that the process was 99 per cent successful as it 'suits my own talents for running a classroom.' However, she added that it doesn't work for everyone. To facilitate the group work, Vicki assigns jobs for each person: recorder, reader, reporter, checker or time keeper. As far as a time frame is concerned, Vicki said that after three weeks, the groups in her classes 'take on a dynamic of their own.' She said that each session is critical in a fifteen week semester. She continued, 'Each person meets his or her responsibility.' The groups in Vicki's class remain together during the entire semester. She models the behavior for the students as she moves from one group to the other. Vicki said, 'The facilitator must ask questions they'd never ask.' Even though no system is perfect, Vicki indicated that cooperative learning works well in her classes.

As the semester progressed, I tried to incorporate some of Vicki's suggestions, and the students responded positively in cooperative groups. Their dedication to the task was noticeably pronounced, and they seemed to respond to the probes about improving the groups more honestly. At this point I wasn't directive about their process unless I noticed that students hadn't accomplished their objectives which I wrote on the board at the beginning of class. However, we established a time limit for the task before they began working in their groups. Sometimes they exceeded the time, and I asked them how much more time they needed.

About mid-point in the semester, I asked students to respond to questions involving their reactions to feedback groups. Out of the seventeen students who responded, sixteen thought that comments from other members of the feedback group were helpful. However, some students indicated that all students should participate and react openly to each of the papers. Another student said that all students should complete the drafts and that everyone should volunteer to read his or her paper. Qualifications included more time, smaller groups, verbal responses, a core of questions concerning the drafts, improved guidelines or directions for each day, and alternating groups. I incorporated their suggestions into my plans especially the time problem and specific directions.

At the end of the semester, I asked some of the same questions I'd been asking all term such as their opinions about working in groups, the purpose, changes in opinions, success or failure of groups, my responsibilities to improve groups, and skills learned. One of the most definitive comments originated from a student who said that group participation helped her 'to have fresh eyes' to point out ways to improve her writing.

Collectively most students appreciated the purpose of the groups even though one person said he was still uncomfortable sharing his writing. Three students disagreed with the concept of feedback groups, and one person deemed the groups successful but not for him personally. However, he recorded that evaluating essays objectively was one of the skills he had learned. The students included other skills such as finding errors quickly, listening, proofreading, keeping an open mind, stating opinions honestly, revising and reading critically, and editing. Judging from the skills listed, the groups were successful for some students. However, the structure, procedures, and debriefing process could be improved as well as the teacher's role.

One student, Amy, said the groups worked well for the students but not for the teacher. She's right. I walked around and listened to comments and made occasional suggestions. However, when a teacher converts to the role of facilitator, awkward moments occur in the transition period. The transition is complete when the students don't need to be told what to do; they're taking control and sharing ideas with other group members.

Improvements for cooperative groups were mentioned by several students. One student suggested that I designate smaller groups. Another student said that I should be more direct in my criticisms and more attentive. The suggestion that group effort should be graded was also mentioned. Some of these ideas will definitely be incorporated into future plans.

Many students learned to consider their immediate actions within the groups and to develop a group consciousness. By evaluating what we did in class, individuals were more aware of themselves as learners. Sharing their reflections with others, either through a compilation of their responses or in large-group discussion, they were more willing to contribute ideas. This concept was also reinforced in the individual conferences as I tried to encourage them to take control of the session by bringing up questions and problems about their papers.

I intend to keep working with students to help them become more aware of their own potential as learners and participants in their own process. If students become immersed in the process, perhaps they'll forget to count the minutes left in their composition class. We'd made considerable progress during the semester, but I wanted to improve group feedback the next time I taught Composition 101, which was spring semester 1994.

Restructuring

I changed my philosophy concerning cooperative groups in August after I attended a Career Track Seminar for professionals entitled 'Implementing Self-Directed Teams.' I finally realized that the group process takes concentrated effort over a long period of time. When I heard the trainer, Lloyd Arnsmeyer, answer questions from individuals in business and industry, I realized their commitment must continue for at least five years. I was expecting my students to assimilate the group process in fifteen weeks. The trainer also emphasized the importance of modeling and providing a means of understanding and implementing the process at the same time.

During the training session, I worked through the activities with Marcy, who manages an office staffed by plastic surgeons, and Doug, who works for the Department of Alcohol and Substance Abuse. I understand how my students feel in these situations. However, they're forced into groups while I willingly paid ninety dollars for the privilege of attending the session.

Control was discussed during the workshop, and I realized about midway through the seminar, that I hadn't relinquished total control. I listened to service personnel, engineers, office workers, and industrial managers discuss the problems involved with giving up control. All the professionals in that room wanted to know the advantages of giving up control; after all, we had managed groups of people with total autonomy for a few years, and we had nothing to replace it but promises. I had many questions as I thought about the application to my composition class. How do I walk the line between director and facilitator? How would I know how much to direct and when to let go? Many questions surfaced from that workshop on building teams.

As the spring semester drew near in 1994, I wanted to improve my effectiveness as a facilitator. I knew I had work to do, so I consulted a colleague who was involved in working with groups. I interviewed Mary Connery, the director of the office of Educational Exchange, who also works with students of all ages in Project Adventure. She reminded me of the necessity to model group behavior. Mary described the purpose of the group as 'a safe, small environment to discuss issues.' She said that conflict resolution may be one of the issues. According to Mary, some obstacles to progress include moving too fast and balancing participation among the members. The positive factors include the attitude of students who are 'more willing to take on a meaningful challenge' within the group than they are in individual settings. All in fifteen weeks; I knew we would work hard this next semester.

I was determined to use some of the students' suggestions and the ideas that other human resources had offered. The day before the semester began, I investigated the room where I was scheduled to teach Composition 101. As I entered Higgins dormitory, I went downstairs and found an extremely large, open lounge with seven round tables and a group of armchairs situated against the wall at the other end. Perfect!

The next day when the students came into the room, they selected their own groups. I talked to them about group dynamics from the first day as one of the other professors had warned me that they wouldn't pay attention since they were seated at small tables. I talked about how important it was to operate efficiently in the large and small group setting; I told them that when I was talking, I wanted them to listen. I also asked them to listen to each other when they worked in groups. On the first day we set up guidelines for group behavior:

- Everyone participates.
- All ideas are accepted.
- Offer only constructive criticism.

- Ask questions.
- Be polite.
- Brainstorm ideas.
- Cooperate with each other by listening.
- Work toward a main goal.
- Use opinions to form better ideas.

Those guidelines became our reference point as we worked in groups and reflected on our successes and failures.

In my interview with Mary Connery, she'd stressed the importance of modeling, building trust, and developing independence. I'd realized that modeling didn't merely refer to standing in front of the class and demonstrating a technique but also exhibiting the type of behavior that is acceptable when they critiqued my writing. I also modeled positive behavior by learning all of their names by the end of the second class. I warned them that I was going to ask them to recite everyone's name at the next class. Even though they were reluctant at first, all of them played the name game during the third class session. Some of them said it made them anxious while others said it was fun. Other community-building activities included reading journals aloud, contributing to a brainstorming session, and discussing topics in the large group. All responses were considered, and no 'put-downs' were allowed. I was very stern when I informed them of this edict because I had learned from experience that negative attitudes would infiltrate and hinder the small group activities.

The first cooperative activity centered around interviewing each other in pairs. The students used brainstorming to formulate the questions which pertained to their histories as readers and writers. As an ice breaker this is quite a standard exercise, but they actually enjoyed talking about their former literacy experiences. Later they wrote about the person they'd interviewed and read the paragraphs aloud in class during the next session. Most of the students didn't hesitate to read; perhaps the informal setting of students sitting around tables helped to promote acceptance and eliminate shyness.

The second group activity was devoted to writing questions about the short story, *Stalking*, by Joyce Carol Oates. Their questions were provocative, and they paid attention to the task. I gave them a time frame, and they adhered to it. As we discussed the story in the large group, I was impressed with their openness. I wrote the following reflection,

Excellent discussion of story — very open and willing to respond — never had this before — what's happening? Some *very sharp* kids in there. I think they'll be fun to work with. I didn't get what I wanted to done, but discussion of story was stimulating. They were into it except for the last few minutes. Should quit earlier.

The students also chose the essays they wanted to read from the text, then they wrote recommendations for them. After they wrote the second draft, they read them aloud. Then the students voted on the top three essays. This approach worked because they selected the options, wrote about them, and decided which ones to read. Choice is an important element for success in a reading and writing class.

After the students participated in groups to discuss possibilities for revising the recommendation, I asked them to write down suggestions for improving the process. They offered the following items:

- Don't be critical of each other.
- Voice opinions more.
- Listen to others.
- Take the initiative.
- Criticize in a nice way.
- No cross talk between groups.
- Work to your potential.

During a conference one of the students had informed me that individuals from another table had ridiculed her group's responses during a recent large group discussion. Therefore, I brought up the idea of eliminating cross talk. I knew different problems would arise, but I wanted them to recognize the crucial ones.

Three weeks into the semester, I scheduled another interview with Mary Connery. We talked about how to achieve some of the guidelines that the students had set up for group participation. Mary offered to conduct a workshop with the students as she had some games that might help us bond together. She also gave me two handouts: 'Stages of Group Development' and 'Analysis of Personal Behavior in Groups.' We discussed the different stages of development which are Orientation, Norm Establishment, Conflict, Productivity, and Termination. It seemed as if we were in the Norm Establishment stage at that time; however, later we seemed to fluctuate between the Conflict and the Productivity stages, especially after I restructured the groups.

The workshop that Mary conducted seemed rather undane at the time because as part of the games for 'bonding' us together, we threw several balls and squeegies in the air along with a rubber chicken. However, we laughed and made mistakes and corrected them together, and no one criticized anyone else. One activity involved a silent circle in which we arranged ourselves in order of our birthdays. I was quite anxious because I was so afraid I'd made a mistake. One of the boys literally took me by the shoulders and led me to my designated place. Since he happened to be a student I'd had to speak to that very morning, I was very surprised about his action. When we debriefed the workshop, I discovered that phase was as important as the activities themselves. The following comments are excerpts from some of students' journal entries:

> I think this [name games] is a good ice breaker so everyone feels more comfortable. Maybe . . . the group will work more efficiently. (Alyson)

Your name is very important to you and when others know it, it makes you feel accepted. Personally I'm a shy person. If someone calls me by my name it makes me feel very good and I seem to fit in better. (Kristy)

She helped us in our communication skills as well as our leadership skills. (Val)

I learned concentration and teamwork. With all those balls flying around you were forced to concentrate and work with the groups together. (Joe G.)

I think Mary was trying [to] take a college classroom situation where there is stress and pressure and relax us by playing games that required us to think . . . Some personalities and other traits came out in people that we haven't seen before. (Melanie)

I . . . could now not be as worried about what the other person thinks when he or she reads my paper because we are more like friends and I won't take any bad criticism personally. (Peter L.)

It taught me how important the group is when reaching a main goal . . . individual is also important . . . help by giving an extra effort. (Larry)

I did get a better understanding of cooperation and working in a group. The exercises . . . required that we all cooperated, paid attention, and participated. And these three things are very important for group work. (Jessica)

It is important to work in familiar surroundings and that should include the people. (Ali)

As I participated in Mary's workshop, I experienced anxiety mixed with relief along with my students. After I read their comments, I realized the benefits of a non-academic activity to slow the process down and to laugh together. Elizabeth Cohen, Professor at the School of Education at Stanford University, advocates training students to cooperate with each other. She says,

Therefore, some developers of cooperative learning strongly recommend team-building or skill-building activities prior to cooperative learning that are designed to develop the prosocial behaviors necessary for cooperation as well as some specific skills for working successfully with others. (1994, p. 26)

When I'm working within the semester time frame of fifteen weeks, I believe that we should engage in these activities at the beginning of the semester and throughout the course as often as time permits.

In the conferences, a few of the students had mentioned the idea of changing groups. After the first assignment was handed in, I rearranged the groups. I attempted to place students who would complement each other in the same group. The students responded favorably after the first feedback session. Of the twenty-two students who were present that day, twenty-one stated that they were comfortable with their new groups. One person said, 'We didn't really do that much but we were somewhat participant and aware.' I was pleasantly surprised, but somewhat puzzled that everyone bonded at the first meeting. The situation changed drastically in at least one group.

Two students began to argue each time their group met. I realize that students tease each other and make comments; however, these two students not only disrupted their group proceedings but the rest of the students as well. I walked over and listened to their conversation at one point and then quietly asked them to listen to each other without comment. They complied for about two minutes. Finally the group developed a working relationship, but the differences remained under the surface. Three of the group members talked to me about the problem. I advised them to listen, discuss, and compromise. Cohen, author of the article 'Making groupwork work,' comments about dissension within the group, 'Disagreement and intellectual conflict are a desirable part of the interaction in a problem-solving group' (1988, p. 13).

About mid-point into the semester, I asked the students to list contributions and requests for improving the groups. A few of the contributions follow: 'to tell the truth,' 'listen to everyone's input,' 'listen to what the paper says,' and 'stay on task and not talk about personal stuff.' Some students stated definite goals for themselves. Improvements include 'not to be shy and tell me what is wrong with my paper and how I can change it,' 'honestly tell me what they think of my paper,' 'not be afraid to be critical,' 'be nice about the corrections,' 'be cooperative and understanding of my work,' and 'honesty and frankness.' We're not perfect yet, but the students are beginning to be responsible and ask for help from other group members.

A week before the end of the semester, the students filled out an inventory concerning their reactions to the group process (see Appendix i, p. 117). Of the twenty-one students who were present in class, nineteen stated that they'd worked with other students in science, math, anthropology, history, and English classes. Other students mentioned sports teams, clubs, friends and family. In response to their reaction to groups, all the students except one said that working with other students was a positive experience. Most of the students responded favorably to the question about the success or failure of groups this semester. However, a few of them qualified their response by stating that they preferred the original group to the new one. Others wrote that they knew many of the people in the class before working with them.

Once more I asked them what benefits they derived. Mark, one of the students who had improved the most, said, 'Enhancing my paper with ideas,' 'Understanding how others write,' and 'Different methods used by others who write.' The last two comments seem similar to me, but Mark evidently differentiated between them in his observations. Another student mentioned gaining 'new options on topics.' Kristy, the student who doesn't like groups, said that she was 'a little more open.'

Another question on the inventory dealt with conflicts. Many of the students stated that no conflicts occurred, but others suggested that when they do happen, the students should listen to each other and reach a compromise. One student said, 'When conflicts occur, there is usually a great deal of discussion. Usually it's between two members and the other two help try to solve the conflict.' In one particular case, that's exactly what happened. According to another student, 'students get upset and refuse to contribute.' Scott stated that I should resolve the conflicts. JoAnn said that 'time was the only real answer.'

In the future I want to encourage students to discuss their papers more openly and honestly. I asked them how I could foster that important component of the feedback process. One student wrote, 'create questions that have to be handed in, to get students to think and evaluate.' Rose advised me, 'don't force it so much — it will come naturally.' An anonymous voice suggested that I give students controversial topics to discuss. One of the most interesting suggestions advised me to 'make the questions they ask more direct and less broad.' Since I ask them to formulate their own questions concerning their drafts, I plan to devise activities to improve the questioning technique. If students are encouraged to offer opinions, the results are always fresh and unprecedented. I certainly will consider Rose's comment in planning and implementing group process activities next semester.

Reflecting

As I reflect upon the 1993 and 1994 composition classes, I realize the structure of the group was probably the most significant difference between the two classes. I don't want to emphasize this issue because it's another complete research project, but the groups in the first semester were self-determined and unwieldy whereas the first groups in the second semester were self-selected and self-directed and contained no more than four students. The setting of the lounge was conducive to group work, and I also made definite stipulations about paying attention, which I conveyed directly to the students. Naturally the students socialized for a few minutes, but they worked on task most of the time.

Stressing trust and respect has also improved the attitudes of group members. I demand respect, and I also emphasize the importance of acceptance of others' opinions, even though they may not agree with the person. This semester, students confided their problems, which were occurring within the group when we conferred. Last semester, I was an outsider some of the time even though I thought I had diverted the control. This semester, I had a better sense about when to take control and when to relinquish it.

The comfort level improved much more quickly than last semester. Some of the students live together in Higgins and know each other outside of class. The sense of community developed faster during the second phase partly because of Mary's workshop and partly because of the individual personalities of the class. I provided more time for interaction in the feedback groups and stressed the idea of learning from each other to a greater extent than in the past. Two groups — the home group and the new group — work because the students learned how to cooperate with people they knew and respected before they interacted with other students.

Unlike the other composition class, students developed conflicts within the groups. Sometimes one person didn't participate or personality conflicts surfaced. I realize that these differences occur naturally in daily interactions with other people; however, students become uncomfortable because they're not used to resolving conflicts. I need to be more conscious of these problems and help students understand that they're part of the group process.

The students' reflections indicated positive reactions to the group process. Many of them stated that their writing had improved because of group participation while others wrote about being more motivated during the small group activities. One student said that reading other students' papers helped him notice his own errors. Scott said, 'I've learned that you must sometimes sacrifice your own needs for the group.' He'd participated on teams so he may have learned to be a team player from sports activities. Commenting about reading aloud in the group, Guy said that he had learned 'to hear my writing by reading to myself.' Finally, an anonymous student said, 'I have learned to work in groups better and even take more of a leadership role sometimes.' The comments were varied; all the reactions except one registered a positive outlook.

Implications

The most influential segment of the entire research project occurred when I attended the Career Track seminar concerning building teams. I was convinced if I interviewed individuals who work in business and industry, I would not only gain insight into the group process but also help the students realize that groupwork isn't limited to a college composition classroom. Therefore, I interviewed Fran Barber, who has been instrumental in training workers to be involved in the team concept at two different plants in Cortland. As Fran discussed her experiences, she emphasized that the main purpose of teams is to build interaction. Team trainers and leaders should show evidence of progress to team participants through graphs and other devices. Training should involve problem solving, brainstorming, conflict resolution, communication skills, and evaluation. As I talked to Fran, I realized the obstacle imposed by the time frame of fifteen weeks.

Even though the workforce at the plant where Fran works consists of older people, the team concept has been adopted. She said, 'Everyone finds a way to fit into the process.' Advantages are numerous; some of them include individual expression, integrated ideas, different points of view, varied methods, cooperative research, risk taking, and group presentations. All of them refer to the positive attributes of the cooperative learning in any classroom. How can I help the students internalize these concepts? Of course, they're the ideal goals, but they may be attainable with proper planning and careful implementation.

If students realized what transpires in team building at the workplace, perhaps they'd be more motivated to improve their group process skills. I asked Debby, who works in an insurance office, to answer questions about the process (see Appendix ii, p. 118). When I asked her what college students should know about working in teams, she said,

> That more and more organizations are moving toward this way of organizing workers so you really should know what it's like. It's very different in that your success depends a lot more upon your ability to get along with and *influence* others than upon your own knowledge of the task or assignment. Your opinions aren't worth much if no one will listen or act upon them. This also tends to be the way the workplace often is, even if you're not in a team. You have to sell your ideas and yourself.

Since the personnel in Debby's office were trained in the team-building concept, she has experienced the process of working in teams.

Inviting speakers into the classroom and interviewing them would give students the impression that groups belong to the real world. Video tapes and audio tapes of presentations and panels may help students realize the enormous possibilities that exist for groupwork. If groups of students investigated some of the places which use the team concept in the community and reported back to the class, they'd be more involved in the discovery process.

In most areas of the United States, educators attempt to limit the influence of industry on policy making in schools. The unions are quite vociferous in their campaign to separate the philosophy of the private sector from the public sector, namely Total Quality Management, from infiltrating educational institutions. However, a middle ground could be achieved as we have so much to learn from each other. I've been invited to join a management association which involves industries and businesses in Central New York. They hold a dinner meeting once a month, and a panel or a speaker gives a presentation. Isn't it about time that we joined forces instead of going our separate ways?

Cooperative learning in the schools and team building in the workplace have some common characteristics as indicated by Figure 7.1.

Conclusions

The original questions concerning benefits, drawbacks, the teacher's role, and working together have been answered by focusing on planning, motivating,

Figure 7.1: Common characteristics of cooperative learning in schools and teambuilding in the workplace

Common Characteristics of Teams and Groups

workplace	*classroom*
trainer	facilitator
accent on communication skills	dependence on communication
cross training	sharing roles in group
techniques used in teaming:	techniques used in groups:
(brainstorming, problem solving,	(listening, responding,
resolving conflicts, evaluation)	questioning, discussing)
monitoring and adjusting	reflecting and assessing
updating staff	asking for feedback
planning	restructuring

Benefits of Teams and Groups

workplace	*classroom*
individual opinions	individual participation
integrated ideas	varied opinions
representative decisions	large and small group decisions
varied techniques	enriched activities
risk taking	risk taking
problem solving	problem solving
improved communication skills	operative communication levels
monitoring process	assessing process
quality product	revised paper

sharing, and reflecting. Even though not all of the answers are positive, I have reached some conclusions. According to our observations, the benefits are numerous. The drawbacks include time on task, reluctance about speaking out, and total participation of group members. However, I realize that a few students would rather work alone, and, therefore, their contributions may be minimal. I haven't resolved the problem about the teacher's role; however, this semester I've felt comfortable enough to sit down at the tables with various groups. A student stated that I should participate more, and I'll consider his advice. I observe the actions of the group and occasionally make comments as I try to be unobtrusive but present. We worked together towards a common goal; one of the students said, 'Group work helps to bring the class together as one.' Everyone has an opportunity to listen, share, discuss, and question.

What did I learn? Of all the advantages of conducting classroom research, maintaining an interest in the students as individuals is the primary purpose. When they respond to a probe in class, I can't wait to read their responses. Before the next class, I compile a list of excerpts from their reflections. Since they're constantly making suggestions, I incorporate some of them into the daily routine of the class. I observe students growing and changing their thinking patterns during the fifteen weeks.

Since the requirements of the composition course are extremely rigorous, classroom research helps me acknowledge different ways to approach an otherwise static course. When students are more involved in their learning, I take my job more seriously. I'm constantly asking questions about improving the techniques of presenting the material: How can I make the lesson more challenging? How can I take individual learning styles into consideration? How can I help them create a more comfortable climate in the large and small groups? How can we reflect more often about the process? How can I measure their productivity?

Classroom research invites the initial questions to grow and expand to create new ones. That's the challenge for future action research.

References

BARTON, F. Interview. 25 March 1994.
BOUGHTON, V. Interview. 5 Sptember 1993.
CONNERY, M. Interview. 7 September 1993.
COHEN, E. (1988) 'Marking groupwork work,' *American Educator*, pp. 10–17.
COHEN, E. (1994) 'Restructuring the classroom: Conditions for productive small groups,' *Review of Educational Research*, pp. 1–35.
GOUGH, E. (1987) 'The key to improving schools: An interview with William Glasser,' *Phi Beta Kappa*, pp. 656–62.
S.B. and MACGREGOR, J. (1992) 'What is collaborative learning?', in MAHER, M. and TINTO, V. *Collaborative Learning: A Sourcebook for Higher Education*, National Center on Postsecondary Teaching, Learning and Assessment.

Appendix i

Composition 101 — Richards Inventory

1. What was your previous experience with cooperative groups? If you were involved in groups previously, what was the purpose of the groups?

2. What was your reaction to groups in the past? If your reaction to working in groups has changed, tell how and why.

3. How well did you work in the group setting during this semester?

4. What benefits did you derive from the group sessions?

5. What would you tell a student who was beginning Composition 101 about groups?

6. In the group, what happened when conflicts occurred?

7. How could I encourage students to be more confortable in their groups?

8. How could I encourage student writers to discuss their papers to a greater extent?

9. What suggestions do you have for improving the group process?

10. What have you personally derived from the cooperative group experience?

Appendix ii

Questionnaire about the Team Process

1. How were you trained for working in teams? Describe the length of time, procedures used in training, and your attitude during the process.

2. As you began to work in teams, what were some of your initial impressions?

3. How would you describe your affiliation with a team now? How often do you meet? To what extent does each person have a role?

4. How are conflicts resolved in your team?

5. What are the advantages of working in teams?

6. What are the disadvantages?

7. What should students know about working in teams?

8 The Never-ending Story: Reflection and Development

Sonia Burnard and Heather Yaxley

ABSTRACT *The most important contribution of action research as part of a school's programme is its ability to stimulate change, ideas and construction. This study focuses on this aspect of action research and is part of a continuing story. It not only reflects the thinking and an active involvement in the education of the children in this school for emotionally and behaviourally disturbed children but also accentuates the possibilities for schools and units working with children who exhibit problematic behaviours. Working with emotionally and behaviourally difficult children can be stressful and staff must be able to sustain the working environment. We show here the regenerative nature of action research and its ability to promote not only research in this area but also ongoing support of staff and school development. Here we show how the adventure of acquiring language for self-exploration is introduced into the behaviourally managed environment of Cooperative Play. The development of this active involvement lends itself to the introduction of the companionship of behavioural and psychotherapeutic approaches.*

Exploring avenues that may show us the ways in which to socialize children with oppositional behaviours into a world in which they will have to participate in a more positive and rewarding manner, is a continuing process. Behaviour in itself is a complex and variable phenomenon and we need to find ways of changing and developing our approaches to children who continually present difficult behaviours. We find that using action research as a vehicle to focus on certain areas for development not only looks to improving the children's control of their own behaviour but also to improve the skills of the adults in dealing with the inappropriate behaviours as well as training up appropriate behaviours. In an educational environment, we see that appropriate behaviours are those that enable learning to take place in all areas, social and academic.

Reactions to situations that are inconsistent and misunderstood by these children are often aggressive and physical. They come to the school with experiences of adults with inconsistent economic, emotional and social life-styles. This establishment therefore, created the experience of Cooperative Play in which children could enter a structured and controlled situation in which adults and children could have positive group experiences. The importance of creating consistency in the children's lives can not be underestimated. The expectation is that children will cooperate, share, create, learn and talk with each other. The adult will be able to take on a variety of roles and feel confident as observer, leader, starter or counsellor. Cooperative Play is a situation which lends itself to induction, training and research. It is continually observed and monitored by staff and children.

More recently it emerged that as the children become more comfortable in the format of Cooperative Play and their ability to conform to group practices improves, our attention is drawn to a new need. If we are to take away those more aggressive forms of communication that the children had, how do we go about replacing a more socially acceptable method? We have the final group meeting when the children have time to show and talk about what they had learned but the information would often have to be coaxed from the child. In this way the apparently easy ability to put ideas forward was not developing in the children as we had thought it would. On reflection, we needed to develop this weak section in our activity. In order to understand how we have reached this important consideration and how we readjusted to answer the problems would best be placed in the context of the development of Co-operative Play itself.

The Concept of Cooperative Play

It has been difficult in the past, from our experience, to look at situations within schools for emotionally and behaviourally disturbed children. The 'closed' classroom does not lend itself to a whole school approach. The child has often failed in the conventional class and is placed again in that potentially negative situation, albeit possibly more contained. In the interests of the child, a more open and constructive environment allows adults and children to share the problems. Clearly, the child must have ways of learning how to learn and how to socialize appropriately. In the best possible world, the educational institution should be as a theatre for its own development and research. The results can only be more beneficial for the child. Cooperative Play came into being initially because we needed to focus our attention on how learning new social skills and teaching skills could come together in a structured environment and make way for a new spirit of sharing; staff and child, child to child, teaching staff to care, management to all staff.

Cooperative Play opens doors, it helps pool information, has a variety of training implications, helps to develop communication and helps children to understand the idea of working together. The structure of Cooperative Play is as follows; three classes of children come together and are mixed into four groups. Each group has a colour and therefore a growing group identity. Ten adults are involved and all have a particular role to play. Termly, roles may change so that adults can practise other skills. The whole 'Play' lasts for an hour and a half and is divided into four sessions. Each session lasts twelve minutes. There is a meeting to start. At the meeting the children are asked to describe what cooperation is all about. Leaders who have prepared an activity describe the activity and the part of the school topic to which it relates. At this time, every leader explains his or her own target and how s/he hopes to achieve it. Each group then chooses which activity it will start at. As they move to that activity, they are given a token. The token is given for listening and talking in turn. All the activities are rewarded in this way by one token. The token is part of the greater school structure which is run on the token economy system. The leader and group members work as a team and the leader joins in as an active member. More recently the groups have an older child from the school who joins as a model for the younger children and as helper and active member of the group. To give them status as helper, they sit on chairs at the meeting and receive their tokens or points, if they have moved onto a point system, in 'bulk' as payment at the end of the 'Play'. Moving to the activity is an important part of the sessions.

Being able to finish one activity, being able to line up, being able to receive positive social praise and a token and to move past other children to the next activity, are all areas of interaction that the children find difficult during any part of the day. They are here put into position to practice and to receive support as they learn. Being an observer alone focuses staff onto the ability to observe and consider children's behaviour and to take this skill to the rest of the job. The bouncer (see below) must have patience and develop ways of talking to children about their strengths and what they have to offer their group. The leaders learn to develop language and social praise techniques. All skills can be generalized to other parts of the school.

The leader is only required to deal with positive interactions, to say good things, to give information and to keep the group working as a team. When a child begins to show that s/he is having problems that are not appropriate — fighting, bad language, throwing, not sharing, leaving the group, for example — then the child is removed by the bouncer.

The role of bouncer is complex and requires a great deal of skill. The bouncer wanders between all the groups, joining in and encouraging but when the bouncer recognizes a child having a problem that may effect the other children and unsettle his own reasonable behaviour to that point, then the child is removed. The bouncer then has the difficult job of finding out about the problem, looking for ways of solving it with the child and preparing the child to go back into the session and start again. This idea is part of the school policy that asks the child to start again and to earn the next token. The child then does not feel that the behaviour has to continue. S/he does not feel negative about the behaviour because the 'next thing' is important. When the children go back to the group, they are welcomed back, made to feel that they have been missed and that they are needed. The bouncer records all removals, the activity removed from and any reasons and for how long.

The observers keep track of group behaviour rather than individual behaviour. Their chart has one minute intervals to record in seat, on task, leader responses to children and talk. They record the number of tokens earned, whether the leaders are achieving his or her target, children removed and they also record some individual achievements for mention at the meeting. The data is kept as part of the ongoing interest in improving the service. Observers' comments have often pointed to more refinements to the system. The ability for children to cooperate within the group is therefore an important part of learning sociability. The observer talks about the positive results and efforts of each group in turn.

Other roles the adults may hold are those of facilitator, 'starter and timer', or observer. The facilitator makes some of the activities a little easier by helping on a practical level if the activity is difficult to manage. The role also requires the presentation of all the results of the work on the walls for the final 'showing' session. The 'starter and timer' makes sure that each session is twelve minutes long and gives a two minute warning so that the groups can prepare for the change. The starter also has to organize the showing at the end. The children need to learn to use their language and to practise talking socially and positively. Therefore an adult rewards the children with tangible rewards for good listening whilst the showing continues. At the end of this session an adult records comments made by the children with regard to the sessions and to put forward ideas or complaints.

All activities are chosen and carefully prepared. They are all based on the school topic so that the information is reinforcing classwork. The tasks are prepared so as to stimulate and motivate the children. They all have end products that are satisfying to the children. The game is constructed just for Cooperative Play and is a popular part of the sessions. It gives opportunities for children to take turns and to use language and to cooperate with each other in short and positive game experiences.

Action Research — A Catalyst for the Early Development of Cooperative Play

In order to understand the more recent developments here, it is important to realize the way that we have arrived at current practice in our action research and how this provides the catalyst for future considerations.

A series of questions arose out of observations and discussions amongst staff and children and these led to the following developments. Questions that arose over a period of time were as follows:

- How do we openly retrain the children's behaviours in order to facilitate their return to traditional structures and situations?

- Could we go further with the organization of the session activities themselves? How could this be addressed for the benefit of staff and children?
- How can we improve our observer charts to give us better data on group responses? Initially the observer charts were primitive in their attempt to look at group social behaviour. They were good enough to prompt staff discussion, identify areas of need and a basis from which to develop and formulate ideas. After a while it was felt that we specifically wanted to look for data to back up what we thought we were finding.
- How could staff be involved in a variety of roles whilst maintaining consistency for the children? The roles of leader, bouncer, observer were originally allocated to staff in an arbitrary way. As Cooperative Play developed some staff asked to change.
- Senior management expressed an interest. What would be the considerations for this move? It would be important for members to see senior management in equal participation.
- How could the session activities be changed so as to improve attention and children's ability to stay with the group? The practical activities on offer to the children throughout the session were looked at more carefully. There was a need to develop the activities to improve the child's attention so there was a need to be more creative with what was given to them. This is particularly relevant with some children with emotional and behavioural difficulties who have varying forms of attention deficit. The activities had to be ready to put the child into a positive situation.
- How could we encourage the children to think about what they were being asked to talk about? The show and tell session at the end of the activity as a whole was loosely organized around the idea that all children should be given the opportunity to show something or to say something. The children would come up as a group and show what they had done rather than say anything about it. As well as the need to develop language skills there was the opportunity to take advantage of the topic bias and therefore the children had more available to them to comment on directly.
- How could we emphasize how important it was for children to listen? The expectation on the children's ability to listen has grown over time without redress or consideration. The children were having to listen throughout the session in order to contribute in the meeting at the end. The activities had become increasingly more complex and this together with the requirement to contribute at the end meant that listening skills had to be actively worked on and reinforced. There had developed an increased opportunity for lack of attention and concentration which needed to be addressed.
- How could we address the needs of children who persistently required

removal? From the bouncers' charts we noticed the isolation of some children who could not integrate with the group and were being removed often. To withdraw a child on a regular basis may be appropriate for an initial period of time particularly when introduced to Cooperative Play but over the weeks this should become less frequent and monitored closely. The strategy was not designed as a solution to disruptive behaviour on a sustainable basis. So for those children who continued to have difficulties which resulted in removal from the group we needed to look for more positive and lasting solutions. We needed to focus on rethinking their needs.

- In raising the interest level of the activities, we recognized that they sometimes required more adult assistance for positive results. How could we include other adults without losing group cohesion? Children were being encouraged to learn new skills and the time was short to get a successful completion of the activity. This in itself created an unwelcome frustration within children and adults to get things finished within the time period. Furthermore, for the child there were potential difficulties arising from not seeing a finished product and therefore perceiving failure. The activities were not specifically designed to increase self-confidence but the effect could have been the reverse in that a reinforcement of failure and lack of self-worth would increase their inappropriate behaviours and subsequent performance.

- How can the non-teaching staff involved be more informed about the increased content of session activities? The staff voiced the feeling that they needed to be more informed regarding the topic. For non-teaching staff their time commitment to Cooperative Play was such that they may come direct from other duties and with the increased expectation and other developments over time they expressed concern that they were ill-prepared for the session activities, both in content and format.

- How could we increase children's involvement in the critical evaluation of Cooperative Play? It became apparent from the data that a lot of information was taken and fed back to the adults but not to the children. We were making assumptions about what they enjoyed and how they perceived the behavioural expectations from the sessions.

- How can we make children more aware of the leaders own commitment to personal change and positive interaction? Discussion revealed that staff were becoming more aware that the behaviour of the children was affected by the behaviour of the adult, as shown by the positive and negative responses of the group leaders as recorded by the observer when correlated to the behaviour of the group.

Developmental Responses

In response, Cooperative Play developed along these lines. We looked to balance the traditional and expected practices with new active change. In this way we sought to build upon what we already had and to maximize good practice through a creative reappraisal of teaching method. In addition to this we wanted to look at how the simple task activity base of Cooperative Play is perceived from the child's position. There was a need to recognize the need of the child in his or her own development by finding a way to make it clear what is in it for them in terms of the perception of Cooperative Play as well as the sum of its component parts.

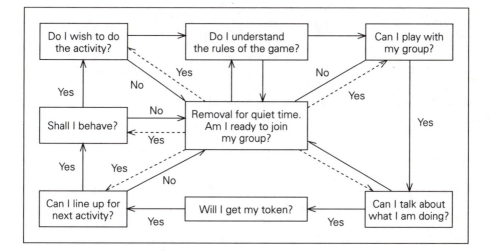

The idea at first was concerned with quite unsophisticated skills, such as drawing, lego, games, which were for the most part unconnected. In order to facilitate meaning within the context of the learning experiences of Cooperative Play we recognized the need to make the connections for the children.

On an initial review of Cooperative Play, staff wanted to link the activities to class topics. It was decided that the activities would fall basically into games, cutting and sticking, modelling and drawing or painting and each activity would take on some aspect of the topic. This would become important because the activities would link into the final meeting where each child would be required to show and tell one aspect of the four activities. Each child would be encouraged to talk to all children and staff with a particular focus in mind when sharing what they had done that morning.

Through discussion and practice, charts were refined and redrafted to focus on particular types of behaviours that were observable and reliable. This in itself is worthy of mention. In the true spirit of action research, the development and drive to initiate change in one area facilitated change and development in another. In this case staff were encouraged through their own perceived necessity to make clear, reliable observations. As an art form in itself this was an interesting and demanding practice for those with no prior experience of data collection, and resulted in the following chart.

OBSERVER SHEET — CO-OPERATIVE PLAY

Date .... Group Colours .... Leader target .... | N/A | A | M |

Leader Target 1 ....

Leader Target 2 ....

GROUP .... GROUP												
Minutes	1	2	3	4	5	6	7	8	9	10	11	12
Activity												
(a) Leader target												
(b) Leader positive												
(c) Leader negative												
(d) Leader silent												
(e) Group sharing												
(f) Group standing/sitting												
(g) Group on task												
Number of children removed												
Number of tokens given for twelve minutes												

GROUP .... GROUP												
Minutes	1	2	3	4	5	6	7	8	9	10	11	12
Activity												
(a) Leader target												
(b) Leader positive												
(c) Leader negative												
(d) Leader silent												
(e) Group sharing												
(f) Group standing/sitting												
(g) Group on task												
Number of children removed												
Number of tokens given for twelve minutes												

We made roles interchangeable. Staff could try any role. We used one review session to see the level at which staff understood the different roles and what skill would be required of them in the fulfilment of a particular role. If adults were going to change roles we needed to have some kind of discussion about what each role expected from each other. Staff now discuss whether they want to change role on a termly basis and this itself motivates examination, change and development. This supports and reinforces the general lack of hierarchy amongst staff within the school in that teachers, child-care staff, classroom support assistants and senior management staff operate at any role. The children are encouraged to perceive the action and role of the particular member of staff at any one time and this is further promotion of the flexible management system which allows children to see staff working collaboratively for the their sake. (See roles below)

Role Definitions

Bouncer's Skills
- Remove children with minimum interruption to group.
- To remain calm when dealing with a child.
- To positively encourage child to return to group for good things as soon as appropriate.
- Ability to recognise developing situations.
- To comment positively to all groups.
- To contribute to and remain in role during feedback.

Leader Skills
- To be a working member of the group.
- To encourage language skills.
- To impart information and direction while the group is in progress.
- To enable all group members to have equal input.
- To reward children socially and with tokens (positive reinforcement).
- To keep ideas of working together as a focus.

Observer Skills
- To be informed enough to understand what makes data valid.
- To remain apart from any group activities.
- Skills involved in recognising group cohesion.
- To be able to differentiate various levels of skills, social and academic, within the group.
- To feed back information positively.
- To develop a sense of time-recording skills.

Starter and Time Keeper
- To prepare children for the session/focus on behaviour and topic.
- To establish each group's new identity and reinforce membership of the groups.
- To time all four sessions and make sure groups/stop/move/start.
- To give all the children the opportunity to show and tell.
- To reward socially and with tokens.
- To oversee the smooth running of the sessions.

Children's Skills
- To remain with group.
- To work to the best of her/his ability.
- To talk to each other.
- To appreciate others' efforts as well as their own.
- To share and take turns.
- Being able to accept direction for starting and finishing.
- To contribute ideas to the group.

Activity Provider Skills

To make sure the activity:

- Is relevant to the topic.
- Can be achieved by all participants in 12 mins.
- Encourages creative and language skills.

Helper Skills (older children)

Model for younger children

- How to discuss and talk to each other.

- How to share a task.
- How to make learning happy.
- Listen to the adult.

Facilitators' Skills

- Moving according to needs of the groups only in terms of their activity and not their behaviour.
- To remain positive and part of the group that s/he joins.

We include management. The Head of Care, the Headteacher and the Head of Education now participate, taking various roles. This not only enables them to retain skills with the children but they can be seen as participating in the ethos of the school. It can be seen as a development for management to maintain social and professional links with staff and children, as well as developing perceptions within the children with regard to staff working together for a common purpose.

Teachers became more creative in the preparation of activities; even engendering a sense of competition and, hence, improving children's attention to group success at activities. The children are now able to approach the coming of Cooperative Play this week with an element of surprise about what they will be able to do. This in turn increases their perceived need to attend and listen carefully at the beginning of the session so that they are well prepared. The intrinsic reward for this attentiveness therefore goes hand in hand with the tangible reward of tokens. Here again the principles of action research as a tool for development are demonstrated.

Children are encouraged to say what they have learned or remembered rather than 'I done this . . .' In this way Cooperative Play fulfils an additional need to explain, providing the opportunity for the child to develop language skills in a structured way. This also meant that the leaders had to be involved in the content of the topic area in order to impart information for the completion of the task, which developed participation skills between the leader and the group as well as between the child and group.

The reward of a jelly-bean was introduced to reinforce verse listening skills at the end meeting of each session. New children are made aware of skills involved in listening and watch to learn the appropriate behaviours of other children who are receiving more jelly-beans. This behaviour can be seen to transfer to other situations, such as assembly times where similar skills have been improved. In conjunction with the school's behavioural philosophy, when tokens and jelly-beans are given the children receive praise and are reminded about why they are receiving them. Staff also learn, as they do this, to look for the positive things and focus on acceptable behaviours.

We offered children a direct model for appropriate and expected positive behaviours. We observed the anti-social and highly disruptive behaviours of one particular child but rather than focus on his inappropriate behaviours we offered him a model for appropriate behaviours. A slightly older and much calmer child befriended him and with adult prompting was able to direct his attention to the target child and focus him within the group. It was then observable that not only did the child's behaviour improve but that the rest of the group were able to function more easily and resume a positive and reinforcing set of behaviours and expectations. The children were also encouraged by the bouncer when they were removed so that they could say something about why they were having difficulties at the end meeting for the children's chart.

We developed the role of a facilitator. The facilitator role was to move from group to group helping out where needed with tasks. This tended to be where the children were involved in the more creative activities and where the help of an adult to 'stick it' or 'hold it' could make all the difference to the child's competence in the activity. To have an adult to carry the precious products from a cutting and sticking activity was a welcome facility for the group leaders too.

Topic information for all participants was requested so that staff had some idea of what the topic activities would involve prior to the session. The adults were then equipped with information, motivating participation, involvement and self-confidence, to impart to the children.

The children are given the spotlight in the meeting at the end of each session in which they can say what they like in relation to Cooperative Play that day. They can criticize as well as comment positively. The information is recorded on a Children's Chart.

The leaders now are required to have their own target and are asked to verbalize it at the beginning meeting of each session. The observer will then look for the behaviour of the leader in relation to that target throughout the session and report back in the open forum at the end meeting of the session. This is in the process of being developed further with staff but is intended to develop a partnership between adults and children which is perceived to be emphasizing the need for all the 'be pro-active' in behaviour management.

Some Discussion of the Data Results from the Above Developments

The children's contributions and the videos show us that children have been learning socially positive language. The bouncer removal charts show that children found it more important to conform to required behaviour than to be removed from the group. The bouncer charts show that for the majority of children being bounced once or twice is enough and that those children then remain in their group. For two children two bounces were needed and for three children four to five bounces were required.

The observer charts show (a) staff are inconsistent in achieving their own targets and in stabilizing a positive approach, (b) negative leader language correlated to low group performance and (c) it was harder to work together as a group when formal in-seat on-task behaviour was required; that task performance was not necessarily undermined by out-of-seat behaviour in less formal activities.

TERMLY RECORD — CO-OPERATIVE PLAY

Date .... Session .... Topic ....

Activity	Green	Red	Yellow	Blue
	1 2 3 4 5	1 2 3 4 5	1 2 3 4 5	1 2 3 4 5
1				
2				
3				
4				

1 Total = (a) .... (b) .... (see Observer sheet)
2 Total = (c) .... (d) .... (see Observer sheet)
3 Total = (e) .... (f) .... (g) .... (see Observer sheet)
4 Total tokens ....
Number of children removed ....

Since we have initiated adult targets it has been obvious over a ten week period that asking adults to review their own behaviours and to target for improvement has been very difficult. One development suggested by a recent observer that may help leader skills would be for all leaders to be observers for a period of time. A recent observer has commented that she found the experience a learning one as she noticed leader behaviours that she felt she had needed to work on.

Our investigations in the past have shown that modelling and creative abilities are lacking in emotionally and behaviourally disturbed children. Rationalization of this observation includes the children's poor communication and language skills and their inability to internalize experience. Using short sessions, with an activity to focus upon and a positive visual result whilst promoting language, maybe a starting point for developing the children's ability.

Cooperative Play compliments, and is complimented by, the collegiate system in the school. We are interested as much in the process as in the results. New staff, parents, governors and visitors can become part of that

process. It is a means of induction into the ethos of the school, positive reinforcement and an active involvement in child development.

Constructive Reflection and New Developments

On reflection it seemed that we had all become good at running Cooperative
Play. Staff were happy to switch to new roles. The starter is no longer a teacher
role as it had seemed to become. Care assistants are equally happy to have a
go at it. The children know when the day had arrived and Cooperative Play
takes its place as part of the children's conversation. But with it appearing to
run well, the weak areas seemed to become more obvious. Because of the
ongoing research situation, these areas are not avoided but rather given a
longer look. The most obvious area is the struggle for the children to communicate at the end meeting. Most children are being prompted to facilitate short-
term memory and are finding it difficult to find the more academic words
related to the topic. The recognition with the difficulty in this area went hand
in hand with the daily need to keep reminding children to use their language
to express themselves rather than other physical modes of expression. The
children show very restricted language and the school looks to extending
children's experiences and language acquisition.

Since the structured environment of Cooperative Play is now part of the
children's vocabulary, they recognize the way it works and what is expected.
New children have its format explained by the other children. They find that
learning to work together is hard and fight against the system initially. However,
it has been found also that a very few numbers of 'bounces' are sufficient before
the children recognize the need to be in the situation rather than to be out. The
rewards of being with the group are greater than not being with the group.

The activities are made inviting and enjoyable and attractive to the children. It seems then, on reflection, that the situation is stable enough to consider introducing a new accent into the 'play'. Those weaknesses suggested
earlier in the format are now very apparent. Observers note that each session
is involved in children learning new language whilst most have not had the
experience of consolidating known vocabulary and structures. So, whilst we
need to be aware of extending vocabulary, we need also to be aware of
children gaining confidence in their own right. Far more obvious is the need
that the children have to be seen and heard in positive ways.

A first consideration then is that children need a way to improve their use
of language and to be able to express themselves in a situation in which they
feel safe. The new consideration then is that the amalgamation of a strong
behavioural structure of Cooperative Play and the token economy system produces a climate in which a more personal language can be developed in the
children .

Secondly, the topic needs to be drawn closer to the experiences of the
children and a situation more akin to a group psychotherapeutic session might

take place within the meeting. Early work with maladjusted children in the 1920s was concerned with providing 'planned environment therapy' and action research was instigated to examine the processes that might improve emotional adjustment. New social experiences and love were advocated. Lessons were voluntary. By and large these failed, possibly due to conflict amongst the adults who clearly would not have enough guidelines to be able to work as a team. Later workers, more notably Anna Freud who made psychoanalytical theory more communicable and relevant for teachers, and Docker-Drysdale at the Mulberry Bush School, provided environments that were largely free of timetabling and had flexible staff functions. A great deal of stress was placed on the individual child and a therapeutic approach was taken to childrens' actions involving destruction and aggression.

It is important to consider this history of approach, albeit briefly, within the last seventy-five years. The environmental approach at this school seems to have reversed previous thought and must therefore take its place in the way educationalists are approaching the children presenting difficulties now. Our environment is created to help the children within a stable and reliable social existence. The children and staff have limits within which to work and a safe place for children to work out their own individual problems. In the end the children must be made as ready as possible to fit into the rest of the world. They have a home, they have a school, they will have work and leisure, there will be behavioural expectations put upon them. With many needs, staff can use approaches within the system that are best suited to the individual needs of the children.

Cooperative Play is therefore a microcosm of this environment but more importantly can be a safe place in which to look for additional practices that involve valueing the child. An important aspect is that of sharing. As Foulkes and Anthony (1984) pointed out to us, this is not a unilateral society. It is important to help and give to children but it is also important to expect something in return. This is a crucial point that is inbuilt. If we speak then we must be listened to. If we have ideas then it is good to share them. If we want our group to be a good group then we all need to contribute. If our leader is working on his or her target to help us, then we can try to achieve ours. Developing language and communication is a high priority and so how would we include this new idea? We have a firm, positive structure and rules that the children understand. Can they take on board the situation as a sounding-board for things that are important to them?

Staff suggested that the best way to help the children to talk would be to prepare themselves in the four sessions and think about themselves in the context of the topic and simultaneously this could give them more confidence when sharing what they know. We had worked very hard on the children's capacity for listening to each other and therefore the situation may give them opportunities to compare those personal thoughts. We needed to look and see through new data whether this would be a viable experience for the children as a vehicle for personal expression, for language development and for sharing.

Methods Used to Watch Development of this Initiative

Suggestions from staff included taking baselines by recording sessions and meetings on video tape. We had previously videoed sessions, but for other reasons. It was also suggested we film a Cooperative Play and have the children comment and talk about what they had seen themselves doing. We could also use it specifically to look at the final meeting and get ideas about how to extend the children's use of language. In the end, however, the tape recorder was suggested for individual sessions so that 'talk' could be recorded. We could also use the recorder perhaps instead of the more invasive video during the meeting to see how the children's preparation had contributed to their ability to share. It is during these sessions, reflecting on past experience and building new ideas, that aptly mirrors Elliott's comments about issues that are part of an authentic action research. The decisions about Cooperative Play are never autonomous, never made in isolation. All our structural changes represent a change in teacher practice and therefore must be acceptable to all those involved. They have common concerns and the children are as involved as the staff (John Elliott, 1989).

The general topic for the school is currently 'Life on Earth' and lends itself to areas that the children could talk about; my family, my friends, my home, parents, relatives, my dislikes, my likes, emotions, and other factors in the child's environment.

What Happened

There were two sessions of Cooperative Play in which we were able to get data for discussion with all staff. However, as an example from one 'Play' the sessions were set out as follows;

Game — The Mask

Children were given oval shapes and felt pens and asked to draw the face of someone in their family and not to tell anyone who it is. They then sat, one at a time, with their mask on pretending to be that person whilst the group tried to find out as much as possible about the person sitting on the chair. The 'winning' group would be the one that could find out the most about these new people in their group in a certain number of minutes.

Drawing — My Room

The children were presented with a 3-D shape of a room and asked to remember details about their room. For instance, did they share, where were the beds, any books, toys or special things?

Modelling — In the Dog House

The children were given balsa shapes, a cube, a triangle and a lolly stick. They were also given squared card, four pins and a pen. They had to make the dog house and put a hooked pin on each side. They could decorate and then write the names of each member of their family on the small pieces of holed card ready to hang on the dog house. They could choose one person to hang in the dog house under its sign written on the lolly stick and stuck on the front of the kennel.

Cutting and Sticking — My Area

The children had to recreate their home environment. They were given various sized squares and strips for roads. They had to think about what their area looked like, where they went to play, shop, visit, and other places they felt were important to them.

What the Children said Behind the Mask

The children enjoyed this game although they had to design the mask very quickly for their role in the game. It was surprising that they took so quickly to the idea of finding out about each other. The hot seat was just as popular as asking the questions, for example:

> *S was his brother* He is older than me. He smokes. He gets them with his pocket money. Mum doesn't know that he smokes. No, he doesn't like school. He wears a pirate earring. He has long yellow hair but he is going to get it cut. He likes rave music.
> *L was his mum* Yes she works. She smokes. She doesn't live with me. She has yellow hair. Yes she has a big nose and mouth. She does cleaning. Yes she loves me. She doesn't live at home because Mum and Dad had a row and because I am mental.

Some children enjoyed being the person behind the mask and others spoke as they would looking at the person from their point of view. One child couldn't think of anyone in his family. He didn't want to be his mother or brother or absent father. So he settled on being a parrot. Of all the activities this game seemed easier for the children as they were directly questioned. The game aspect made it more acceptable to them.

How Did They Manage in Other Sessions

Making the dog house was fun. The idea was simple and writing names from the family and the sign 'Who is in the dog house?' was for most children a chance to place who they felt deserved to be kennelled. For others it presented different kinds of problems, like distinguishing adults from home and school. It was making significant, the relationships they had beyond those in which they were immediately involved.

The children had most difficulty trying to place their homes. Three out of five children could not visualize their neighbourhoods. Those that could enjoyed verbally describing where they went to get to parks and shops. However, orientation on paper was very difficult. For the most part, they exhibited very little information about where they lived and what transport was available or streets that led to places of significance. They had difficulty remembering the colour of their front door, neighbours or where they might go to play.

The 3-D aspect of the drawing confused the children. Most of them couldn't visualize walking through the door into their rooms and what they could see. There were many adult prompts in every group, asking children to remember where their beds were, for example, trying to focus children with more specific questions like 'duvet cover'. Some could only see where they were at school in their bedroom. It gave the adults a feeling that the children had two separate existences and they could not see from one to the other.

What Did They Say at the End?

Observation charts showed that compared to previous weeks, children showed more fluctuations in their group behaviour. Children's responses to showing and sharing varied but in these two last Cooperative Plays compared to previous 'Plays', there was some evidence of difference in the way the children presented but showed little evidence of extension of language or an ability to communicate without adult prompts. However, for many who were less able the extent of contribution was greater. One child explained his room in great detail, even to the spot where the cat who died used to lie. Another could describe his environment very well and showed how his foster parents took him to the park. There were also some anecdotes, like the boys who fought battles with their parents to get their beds in the same room to share. There was also a great deal of sharing of experiences, in that children copied anecdotes from children who had shown earlier in the session and took them on as their own. The tape showed children taking greater interest in saying that their rooms had CDs, mini hi-fis and televisions than describing favourite toys or books.

The tape is able to recall more than one's own recollections at the time: there is evidence that some children responded well and that more talking took place than was apparent at the time. There was almost too much 'about themselves' to choose to think about and say. The activities in themselves required a great deal of thought let alone the different kind of thinking about themselves. What they had to show, at times, seemed to conflict with their own comments of 'what' it was all about.

Staff Reflection

One of the most interesting comments from all staff was how much most of the children cut out thoughts about their home environment when they came to school. For whatever reason they found it hard work to talk about home. This specific social language is very difficult for children with emotional and behavioural difficulties. The children exhibited far more disturbing behaviour during this 'Play' than in others, as reported by leaders and observation sheets. Removal of children went up by two per session according to the bouncer sheets. Staff commented that the tasks were too hard and required too much thinking on the children's part. The children were reluctant to talk about themselves although staff commented that perhaps they hadn't had the practice as yet or the language to deal with the kinds of activities they were being asked to participate in.

Bedrooms or 'own rooms' were not presented as happy places. Only one child talked about toys in the bedroom or represented favourite toys or a stuffed Teddy. Many children said they had televisions in the room and a few that they had a computer. Some couldn't remember if they had windows and none could remember what they could see from their windows. There was only a little talk about posters on the wall and no child seemed to have been involved in colour schemes or organization of the room. Staff commented that children didn't seem to 'know' their environments. A child in one group seemed to recognize immediately that the situation was to be about his central life concerns. His quick appraisal seemed to warn him of the imminent threat to his normal 'play' coping strategies. As suggested by Lazarus (1980), his first response was a denial. He started his bedroom and then scribbled in black crayon all round his picture but also wrote his name on the back. However, when he got to the third session and was reappraising the situation, he was then able to identify his own level of emotional participation and coped with the two sessions at the end by cooperating minimally but talking constructively. It is apparent from talking to all staff that many children used a variety of coping strategies for these sessions.

The general feeling from most participants was that two Cooperative Plays on this subject was too difficult for the children. We had 'leapt' in thinking that the 'safe' structure was sufficient to give the children the confidence to participate. On reflection, most of the changes we had made had been made slowly and over time and assimilated into the situation. Linking home and school is important for the children but must be approached more slowly with greater preparation. We all thought therefore, that the inclusion into Cooperative Play should be into one activity session out of the four. The chosen activity would be arranged after talk amongst all adults and should be linked to the topic and the child.

We also thought that we should be introducing children to a more open language about home at other times in the school day and more consciously with the Keyworkers. It is not sufficient to offer an opportunity to talk about things children know about, they also need to know how to talk and what words to use. However the experience has opened new areas of thought and talk for staff with more definite proposals to be discussed in the general staff meeting. Clearly all children have different needs and various levels of understanding about who and where they are. We need to think about giving the children words to describe their own situation, to understand it from their own focal point and to express themselves. Many children have physical defence and attack strategies and if we are to take those away, they must have something else to replace them.

Summary

Not all developments can be seen initially as successful and this piece of action research needs to be redesigned and repeated. Whilst we set out with a hypothesis that our structure was firm and secure for the children to express themselves with confidence, we found other areas that need work and a whole new attitude toward how to introduce new ideas. The children will need just as much practice in thinking and expressing themselves, whether in topic work or about themselves. These two areas can work together side by side but need equal input. One works to enlarge understanding and vocabulary about the general world and the other to use the information and ability to remember and share at a more personal level. These developments can take place within Cooperative Play but also need to have more consciously specific time elsewhere.

More importantly, the research focused all our attention on the movement of ideas and developments within the school. Cooperative Play demands a great deal of energy and preparation for staff to make it work for the children. It also demands a great deal from the children. Over the year it can have an effect on many other areas of the school. Children know that it is scheduled into the working week. It tells children about how to work on working in a group. It shows and helps children to think about listening to each other which the children must do in assemblies and in classwork. It is also a time when all staff work together and therefore have a joint concern.

Why, then, continually focus action research on Cooperative Play? Possibly because it is a time when the behaviour of staff and children can be discussed openly and difficulties can be shared. More particularly it indicates to staff the regenerative powers of re-thinking situations and developing them. It is this regenerative nature of action research that helps the adults to learn how to have a healthy attitude toward coping with very difficult children and redefining ways of helping them in all other areas of the school. It is an ongoing process of all the staff seeking to improve the quality of practice in this area of education.

One might get into other conversations about this piece of action research. It may be that with the western idea of the promotion of self out of the context of social, cultural and familial situations might challenge the need for the children to place themselves in contexts. We could discuss the importance of social interaction in the development of self-concept, however, this would again be part of our action research in future discussions. The ideas open up areas for discussion and consideration.

We need to continually think about how we help children with all kinds of emotional and behavioural difficulties. It is an area that grows and will need continued exploration. We have the children for only a short time and we need to give them educational and social skills to enable them to live positive adult lives. Action research is built into the system, has many functions and regenerates interest, development and training incentives. This example is one that started with Elliott's (1991) first characteristic feature of the reform process: 'It is a process which is initiated by practising teachers in response to a particular practical situation they confront' and has moved on to explore endless avenues.

References

ELLIOTT, J. (1989) *Studying the School Curriculum through Insider Research: Some Dilemmas*, Norwich, University of East Anglia.

ELLIOTT, J. (1991) *Action Research for Educational Change*, Milton Keynes, Open University Press, p. 9.

ELLIOTT, J. (1994) 'The teacher's role in curriculum development: An unresolved issue in English attempts at curriculum reform', *Journal of Curriculum Studies*, **2** (1), pp. 43–69.

FOULKES, S.H. and ANTHONY, E.J. (1984) *Group Psychotherapy: The Psychoanalytical Approach*, London, Karnac.

LAZARUS, R. (1980) 'Thoughts on the relations between cognition and emotion', *American Psychologist*, **37** (10), pp. 19–24.

Part III

Action Research and Individual Learning

9   Can Action Research Give Coherence to the School-based Learning of Experiences of Students?

Anne Edwards

Introduction

The Grail of the reflective practitioner now appears to have been replaced in programmes of initial teacher training (ITT) in England and Wales by the pursuit of the competent teacher (Maguire, 1995). Arguably this new emphasis complements the move to an increase in the school-based elements of the initial training process and an increased role for classroom practitioners as mentors of student teachers through their involvement as partners in the training process. At first glance, a coherent picture of an apprenticeship model of learning to teach would appear to be being presented. However the idea of apprenticeship deserves to be examined before the package of ITT as a school-based mentored training for competence is accepted as an unproblematic response to what is regarded to be the need to train teachers in the settings that will become their workplaces.

Current neo-Vygotskian perspectives on how learners acquire understandings of publicly valued knowledge in any domain, be it mathematics, weaving or pedagogy, emphasize the mediating role of more expert members of the knowledge community (Tharp and Gallimore, 1988; Wertsch, 1985). Mediation can occur in a variety of ways. These might include the strategic modelling of appropriate actions, the carefully phased resourcing of the novice's activity in order to limit opportunities for failure and, perhaps most importantly, conversations which serve the purpose of inducting a learner into the discourse in which the knowledge base of the domain is encoded or carried. Apprenticeship in a neo-Vygotskian framework is therefore far more than placing a learner alongside an expert. It requires of the expert an explicit understanding of the knowledge base that is to be acquired by the learner and a view of learning that recognizes the role of the teacher as someone who controls the learning environment in order to provide support when it is needed and to withdraw when the learner wants to act independently and is likely to succeed in the attempt.

Learning to teach is a particularly complex process which does not consist of the simple acquisition of, for example, classroom management skills and subject knowledge (Shulman, 1987). It cannot be equated with learning to weave however beguiling a simple model of apprenticeship might be. Classroom teaching comprises sets of rapid decisions and anticipations that are frequently difficult to articulate in conversation with others (Elbaz, 1990). In addition, the idea of *being a teacher* is strongly embedded in students' self-images before they start training. Consequently students arrive with clear pictures of what kind of teacher they want to be and are reluctant to shift those images as they progress through their training, preferring rather to use strategies associated with these images in order to ensure their classroom survival. This phenomenon could perhaps be summarized as the 'Fisher-Price Effect' as the influence of playing at being a teacher with toys like the Fisher-Price Playschool and fantasizing about teaching as telling and keeping order are carried with beginning teachers from their own childhood games and observations.

Perhaps of most relevance to an understanding of how an apprenticeship model might need to be adapted in order to operate in ITT is the need to recognize the public arena in which the beginning teacher has to acquire the skills of teaching. He or she has to succeed first time and has to maintain sufficient self-confidence to return for the next teaching session. Visible apprenticeship to the more expert teacher is likely to undermine both their own images of themselves as coping teachers and their status with pupils. In addition it would seem that certainty is of more value to student teachers than the tentativeness perhaps required of reflective practitioners.

Given the context just described it is unsurprising that the development of the informed reflective practitioner has been a thwarted aim in so many ITT programmes and that with a few exceptions, for example Korthagen and Wubbels (1995), teacher educators are unable to demonstrate a well-substantiated case for its general effectiveness in the initial preparation of teachers. That this situation obtains — despite a heart-felt belief of the value of producing in beginning teachers what MacKinnon and Erickson (1992) describe as 'particular dispositions for enquiry, ways of seeing, critiquing and acting in classrooms' — is perhaps an indictment of the way that reflection on practice has been integrated into programmes for beginning teachers.

In this chapter I propose to examine how the preparation for lifelong reflective practice, which may at times emerge as critical enquiry (Ruddock, 1992), could occur. I shall suggest that it will be necessary to take one step back from a view of reflective practice in ITT as *not quite action research* to see it as essential to the learning processes of student teachers. Schön's work is frequently invoked in discussions of learning through reflection (Schön, 1983, 1987). But, like Eraut (1995), I have long harbored doubts about the relevance of much of Schön's work to the preparation of student teachers in busy classrooms. Instead I want to consider the value of reflection on practice to students as learners. Consequently I intend to regard the frameworks for reflection on action offered in the action research literature as a vehicle for that learning. In doing so I shall draw heavily on neo-Vygotskian perspectives on teaching and learning and in particular on the importance of well-managed dialogues to both informed reflective practice and to the induction of novice teachers into the public discourse of teaching and learning. In this endeavour I shall draw on data collected in primary schools in a three year study of a pilot school–university partnership degree programme which has the creation of reflective practitioners as one of its aims (Collison and Edwards, 1994; Edwards and Collison, 1995, in press).

MacKinnon and Erickson's (1992) notion of 'dispositions for enquiry' is a useful step forward and away from previous attempts at making a case for the development of reflective practitioners in ITT. It places the endeavour as a *within the student* problem and ties it to a focus on the learning processes of the students themselves. Other discussions of reflection and ITT have tended to examine the difficulties involved in transferring current understanding of reflection on practice and critical action research gathered from the experiences of qualified teachers to the concerns of student teachers. Here I would include McIntyre (1993) and Ruddock (1992). While both writers clearly have high quality ITT as a major aim in their writing and in their practice, the discourse of reflective practice in which they examine reflection in ITT is restricted by its development in the practice of qualified teachers.

Faced with the apparent inability of ITT practitioners, in the UK at least, to defend the *reflective practitioner* from the onslaught of the *competent teacher* it may seem timely that the place of reflection in ITT is reconsidered. It would seem important that these reconsiderations occur in ways that might ensure that reflection is relevant to models of ITT that are to emerge from reconceptualizations of teacher training derived from an increase in training responsibilities of school-based teacher-mentors. To do that we need to examine the relationships that might exist between reflection on practice and induction into the professional knowledge base of teaching.

A Neo-Vygotskian View of ITT

The introduction of vocational degree courses for prospective UK teachers in the 1960s was not matched by a coherent view of the phased learning of students on the programmes that were offered. Consequently a variety of structures emerged and were encouraged to flourish even after the Council for the Accreditation of Teacher Education (CATE) took control of the content of the ITT curriculum in the mid 1980s. In a way that paralleled the later development of the UK National Curriculum for schools, the CATE criteria for ITT identified matters of content but left pedagogical processes as they related to the learning of student teachers to the Institutions of Higher Education (IHE). Later pronouncements from central government (for example DfE, 1993) identified the relative responsibilities of the sites of training whether they were schools or IHE but again did not indicate a view of how students learn that extended beyond the need for them to be able to transmit the subject knowledge that comprised the National Curriculum.

Arguably the changes in provision that were to be derived from school–
HE training partnerships should have occasioned a radical review of the learning cycle of student teachers by the two partners responsible for supporting
that learning. Evidence to date suggests that has not occurred (Edwards and
Collison, 1995). Rather, schools are operating as safe places for student trial
and error learning and are glad to be at last appreciated for offering those
opportunities. School-based mentors do not appear to see their role with students to be part of a complex and long-term process of student learning
(McIntyre, 1995). Students who wish to attack their teaching with certainty are
encouraged to do so and evaluative dialogues that extend student understanding of general principles are extremely rare (Collison and Edwards, 1994).

Yet it would seem wise that partnerships in training were premised in a
shared understanding of how students learn and behave in the variety of settings in which their learning occurs. Perhaps even more important is an understanding of how the relative responsibilities of school-based and HE tutors
combine to provide a coherent programme for student teachers that allows
students access to the support they require when they need it. To achieve this
aim all the partners in the ITT process need to share a coherent view of how
student teachers acquire professional knowledge and what it is that particular
settings might offer students as they progress towards an understanding of the
complexities of teaching.

The need for a coherent pedagogy that can give direction to ITT has been
recognized. It resonates through Bennett and Carre's (1993) account of a
postgraduate training programme in one English university. It is also tentatively tackled in Edwards (in press) as a learning cycle that, in neo-Vygotskian
terms, moves the learner from the introduction to ideas and experiences in the
public arena of, for example, an academic text or lecture to the consolidation
of personally held knowledge gained through experience in the more private
arena of a mentoring conversation or tutorial and finally out to the public
arena for competent performance and critical enquiry.

A neo-Vygotskian view of student teacher learning would therefore see
the learning process to consist of at least three stages. In the first stage, student
teachers are introduced to key concepts in an aspect of professional knowledge while in an IHE or while observing in a school but at that point they
would have only a partial grasp or even a misapprehension of the ideas. That
would remain the case until they had tried to work with the ideas in their own
practice and had talked about them. This opportunity would occur in the
second stage when students are immersed in the demands of practice in school,
and it is here that they grope towards personal understandings of public knowledge associated with teaching. The final stage in the learning process would
come when, confident in their mastery of that element of the discourse of
teaching, students can draw on their own informed observation to assert, in
Whitehead's (1985) terms, their own claims to knowledge and offer a critique
of, for example, current practice.

The stages of knowledge acquisition just outlined parallel those used by teachers in schools as they first move children through tasks that allow exploration and introduction to the language of the subject. They then move on to structured group activities, where they can test their new understanding and use of the subject discourse, before finally progressing to public performance, by applying the new understandings in complex problem solving tasks, or by simply improving the speed and accuracy of the performance. The most difficult stage for the teacher in this process is the management of the second phase where the learner has to begin to use the discourse of the subject. While language support is frequently necessary at that point, the teacher has to provide it in ways that do not revert the learners to the dependency that Prisk (1987) describes in her account of teacher intervention in group activities in infant classrooms.

In the student teacher learning cycle the second phase occurs most frequently in school settings where support is to be provided by mentors. Our data reveal that while the mentors who worked with us in the study were confident that they could give the well-resourced context for practical learning they felt that it was not appropriate that they should challenge the students about their practice. Furthermore they stated that matters they described as 'theory', i.e., not directly related to immediate task management, should be addressed by the higher education partner. Consequently although students were gaining access to ways in which they might work with, for example, a historical artifact to encourage pupils to engage with the discourse of history, they were not finding themselves in conversations with teacher mentors that enabled them to engage with the discourse of pedagogy (Edwards and Collison, in press). The pupil-centredness of the primary school classrooms in which the student teachers were operating appeared to be militating against attention to the need to support students as they too were inducted into being users of the body of knowledge which is arguably at the core of being a professional teacher.

An analysis of learning as a three stage cycle allows us to see the relationship that can be obtained between IHEs and schools as both sets of institutions endeavour to support the learning of student teachers. Such an analysis may allow us to see that each partner may take the lead in supporting students at a particular stage. It was clear from our study of primary school mentoring that the mentors saw their prime purpose to be to provide safe places for students to practice the skills of teaching. The teachers wanted to avoid risks of student failure, consequently they carefully supported or scaffolded the learning contexts in which students found themselves by focusing their conversations with students on how to resource and manage learning tasks (Collison and Edwards, 1994).

As part of the study of mentoring in primary schools five hours of mentor and student conversations were recorded by participants. Eleven teachers and twenty students were involved in recording conversations over one year of the degree programme. The majority of the interactions focused on planning for future sessions. Analysis of the talk of the mentors to student teachers revealed that 45 per cent of the units of talk were instructions related to managing tasks in the next teaching session, for example, 'you'll need to make sure you have the paper ready'. Twenty one per cent were points of clarification, 'it's in the cupboard'. In 15 per cent of the units of talk the mentors indicated that they were listening to students' comments and were building on them, 'that's a good idea, why don't you give them the words too'. Only 7 per cent of the total number of units of teacher talk were considered to be challenging students to think for a second time about the actions they had taken. The remaining 12 per cent included references to how children learn, relationships with parents and discipline. It was evident from these data that an emphasis on competent performance was meshing with the students' desire to be teachers, and the mentor's concern to avoid disruptive sessions which would both unsettle the pupils and unnerve the students. As a result, reflection took second place to instruction. This situation was exacerbated by the relative dearth of conversations that were evaluatory in intent.

The picture presented here, drawn from analysis of mentoring conversations, questionnaires and interviews with teachers and student teachers (Collison and Edwards, 1994), raises a question about what students were learning while in school. These questions are not new (McIntyre, 1995).

Dialogues in Teaching and Learning

The picture of the experiences of students in school-based ITT from the three year study suggests that, as in other programmes, opportunities for their learning while in school were not being maximized. A major weakness in the student experience identified in our data was the lack of evaluatory conversations that could provide the basis for creating a 'disposition for enquiry'. Observational data in the same classrooms helped to provide some explanation for the lack of reflective analysis. These revealed that the vast majority of mentors were regarding the students as useful pairs of hands in classrooms which made it possible for mentors to attend to the pressing needs of specific groups of children while students worked with the remaining children (Collison, 1994). There were of course noticeable exceptions to this practice but those mentors who watched and supported the student teachers in their classrooms were not particularly appreciated by the student teachers who were assigned to them. These students felt that they were not being trusted to be teachers.

The importance of privacy from the gaze of other adults has been well documented (Tickle, 1993). One consequence of the strong desire for privacy while teaching is that scaffolding conversations between mentors and students that build on shared experiences are difficult to achieve. Dialogues that might have supported or scaffolded the learning of students while engaged in the practice of teaching or immediately after the session therefore could rarely draw upon a shared database of observations of pupil or student behaviour.

It is at this point that I would wish to propose a view of reflective practice that recognizes that a central aim of reflection on practice is to promote the individual learning of student teachers. Again this is not a new proposition. It is evident in, for example, MacKinnon and Erickson's 1992 analysis of the relationship between student teacher development and reflective practice. Here they argue that a constructionist perspective on student teacher development has to build upon previous understandings and 'the ways in which these interact with their current observations and interpretations'. It is, as has already been argued, the management of this stage in the learning cycle of students that is so tricky. Mentors need to acquire the right to examine the details of student practice as well as have access to previous ways of understanding particular phenomena held by students. Yet they need to be able to do it in ways that do not undermine student confidence but, conversely, enable students as they are inducted into ways of talking knowledgeably about professional practice. Current emphases on classroom performance in ITT in the UK appear to have produced some alarming side effects captured in fieldnote data. For example: 'We did evaluation on the last practice (i.e., we don't need to do it in this one).' It would seem that ways of counteracting this effect are urgently required.

Griffiths and Tann (1992) identified five levels of reflective practice that may help to provide a framework for understanding how reflection on practice might be built into a coherent view of how student teachers acquire and use professional knowledge. Their levels of reflection, which they are at pains to emphasize do not form a hierarchy, comprise two levels of reflection in action and three levels of reflection on action. The first two they describe as *act – react* and *react – monitor – react/rework – plan – act*; the final three are review, i.e., *act – observe – analyse – evaluate – plan – act*; research, i.e., *act – observe systematically – analyse rigorously – evaluate – plan – act*; and retheorizing and reformulating, i.e., *act – observe systematically – analyse rigorously – evaluate – retheorize – plan – act.*

In the descriptors provided by Griffiths and Tann the difference between reflection *in* action and reflection *on* action appears to lie in the extent to which there is the opportunity to take the time necessary to think and in the extent to which, in the final two levels at least, that reflection is informed by public knowledge. The time dimension is made much of by Eraut (1995) in his critique of Schön's influence on teacher education. At this point I would prefer to focus on the power of well-managed reflective conversations between mentors and students to bring together the elements of the student experience that may be regarded as theory and practice. We found very few examples of reflective conversations in our data and those that did occur were more easily placed in Griffiths and Tann's (1992) category of review than in research or retheorizing and reformulation. Nevertheless a few mentors were encouraging the 'dispositions for enquiry' that would seem an essential prerequisite of the continuing professional development of the student teachers.

The following extract is from a conversation between a student teacher and her teacher-mentor in an infant school. The extract starts with the classteacher-mentor (*CT*) picking up on the evident pleasure of the student teacher (*ST*) after an account of a successful activity in the previous session which was meant to have had a religious education (RE) focus.

CT So you were pleased with it. Would you do it again?

ST Yes, but maybe not with Derek.

CT So how would you adapt it for him?

ST Well I think that it would have to be brought in more slowly, in stages now, you know starting with like the paints because obviously you have not had as much experience with the palette as the other children have. If you just spend time with the palette at one stage.

CT Well that's correct but . . . so your activities . . .

ST Yes but he did it.

CT So with him what would you do the next time? I mean, say for instance you have children who haven't had painting experience, how would you deal with it?

ST Well I think that I would have given them the paints to start with and let them have a session just mixing angry and sad colours and then moving on to a discussion.

CT Yes, thinking about, because you are coming on to the science, the arty side. Thinking about your RE, could you have done the activity with the children in another way?

Here the mentor is using her dialogue with the student to encourage student confidence (note all the affirmative openings) while at the same time she is trying to allow the student to become her own critic, while simultaneously attempting to keep a grip on what was meant to be a specific curricular experience for both the pupils and the student teacher.

This extract cannot easily be equated to the type of conversation that might occur between a practitioner who is an action researcher and his or her critical friend. While the teacher is using the conversation to help the student make sense of her experiences there are some vital missing ingredients for this to be a *real* conversation about a piece of past practice. Its unreality lies in the fact that it is an example of gentle didactics in a situation where the person doing the teaching knows less about the actual event than the learner but is nonetheless the expert. The missing ingredients therefore include a set of goals against which the student's teaching task may be evaluated by both mentor and student and some evidence on which that evaluation might be made.

We found no examples at all in our study of mentors and students identifying either pupil or student learning goals prior to an activity (Edwards and Collison, in press). As a consequence, evaluation of task or student performance against identified goals was also absent from our data. A review process in the definition provided by Griffiths and Tann (1992) was taking place in the extract just presented but explicit references to pupil or student learning goals and the evidence on which actions were being evaluated were absent.

Also missing, if we are to return to MacKinnon and Erickson's (1992) constructionist perspective on student learning, was the mentor's access to the student's previous construction of how children learn and how they particularly acquire key concepts in RE. Consequently, despite all her skills as an attentive listener who is taking seriously the student teacher's interpretation of events, she is not really able to present the student with a way of taking her thinking forward that may itself be tested in practice and evaluated. The encounter lacked a structure that would give the student the opportunity to incorporate systematically her mentor's critique into a way of firstly reconceptualizing the task under discussion (Korthagen and Wubbels, 1995) and subsequently moving on to Griffiths and Tann's (1992) research level where a revised perspective on children's learning in RE might be informed by additional reading. As it is there remains a nagging doubt as to whether the student gained anything at all from the thoughtfully framed guidance that the mentor attempted to provide.

Public and Private

The outline of what has been described as a neo-Vygotskian model of teaching and learning makes a distinction between what Vygotsky refers to as two planes of learning (Vygotsky, 1978). The first encountered plane is the public arena in which learners are initially acquainted with new information and eventually, after acquiring a personal grasp of the key concepts and skills, come to demonstrate competence in their use in publicly valued ways. The second plane is the private arena in which learners grope towards understandings by tentatively testing out their new ideas and the language in which the ideas are carried. It has already been argued elsewhere that the framework provided by a Vygotskian view of learning as progress from the intermental (public) arena into the intramental (private) arena and back out to the intermental is compatible with understandings of the value of action research to supporting the learning of qualified teachers (Edwards and Brunton, 1993). The argument runs that the (inservice) tutor who takes on the role of critical friend provides a bridge that allows the enquiring teacher to move from the public knowledge domain of discussions of practice to the comfortably private arena of his or her own classroom and out again into what one of the teachers in Miller's (1990) examination of an action research inservice experience described as 'the wider world'. As the teacher moves from his or her classroom to a more public discussion of events the critical friend creates a context in which the evidence brought from the classroom of the enquiring teacher is respected and is allowed to become a basis for an informed discussion of the preoccupations of the teacher. These preoccupations have often largely been identified before the action being evaluated was taken, though of course the focus of enquiry may change radically as the enquiry proceeds. The crucial feature of the evaluation interaction between presenter and critical friend is the evidence that is offered as the basis for joint discussion.

In the extract from the mentoring conversation that was discussed earlier, it is this shared starting point that is missing. The student had not identified a particular area of practice on which to focus her attention and had not gathered evidence that would have given her something that could have been explored in detail with her mentor. The conversation operated at the local level of a particular class and a particular child and although the mentor tried to lift it to a more general level she was unable to do so. Small extracts of, for example, pupil talk or work can, as we know from experience of action research conversations, open up so many possibilities for discussion of general pedagogic principles. Above all they do so in a way that turns the conversation from a justification of previous practice into a joint exploration of an interesting piece of data from a classroom through which all the participants in the conversation can tentatively explore their understandings of the situation and bring into use the language of pedagogy.

One of the difficulties inherent in talking about ITT is the language collision that occurs when trying to think of the student teacher as learner. The learners in classrooms are the pupils. Classrooms and curricula are designed to that end. Pupils need semi-private places, often classroom groupings, in which to try out safely their recently acquired subject-specific language (Edwards, 1994). Classrooms cannot operate in the same way for student teachers. Classrooms are places of public performance for beginning teachers, hence their desire to limit the audience to those who they perceive will be least critical (Tickle, 1993). The second phase in the learning cycle outlined earlier has to be specifically created as an unthreatening place where students can acquire and use the professional discourse of teaching in tentative ways. They cannot do this while teaching pupils in classrooms.

Miller (1990), talking of inservice teachers, advocates the creation of 'spaces where dialogues can occur'. This telling phrase brings together some important features of what might comprise a useful way of looking at how mentors might support students as they learn from their experiences in classrooms. First of all the conversation between mentor and student has to be seen as the place where practical learning is subtly categorized and consolidated in ways that are compatible with current understanding of what makes for good practice. Without that process of categorization and consolidation with an expert there is always the danger of the dominance of powerful misconstructions of events that might inhibit the future learning of a student. Consequently conversations that draw on recent events have to be a priority. Our data indicate, however, that discussion of previous sessions were considerably less frequent and shorter than were planning conversations. In addition, even fewer of these discussions could be said to have been managed in ways that encouraged students into 'dispositions for enquiry'.

The conversations that it is being suggested should occur are best seen as dialogues in which student teachers are able to try out the language they need to explain or hypothesize from their own experiences as teachers in classrooms. Shotter (1993) describes the role of the instructor or teacher in a Vygotskian model of teaching and learning to be,

> to 'stage manage' the context of 'joint action' which 'calls out' what in some sense we can already do, and this helps us to recognise how to call it out for ourselves. (p. 94)

Shotter is making reference to Vygotsky's view that one can only control a mental function, that is categorize and use an idea, once it has been used unconsciously (Vygotsky, 1962). Here we can see that, from a Vygotskian perspective, classroom practice without a mentoring conversation that 'calls out' what is relevant is unlikely to be learning.

Beginning teachers will rarely have the focused eye that will enable them to select what is relevant from the vast array of activities that comprise a teaching session. At the same time it is unlikely that they will display mastery of the pedagogical discourse they are to be encouraged to use. Consequently mentors have to manage the dialogues in ways that induct the student teachers into the concepts they are to acquire and 'call out' the language in which these concepts are carried.

Much can be learnt from studies of early language acquisition (Bruner, 1983). From developmental psychology we discover that young children are inducted into their particular cultural backgrounds through carefully managed conversations with older members of the community. At first the more expert acts as if an infant understands what is being said and the modes of appropriate response. Eventually young learners begin to use the language formats and conversational strategies offered to take control of the conversations. The first signs of this are, for example, when a child starts to play games with or tease the adult with whom they are interacting.

If spaces for dialogues are to be created for mentors and their student teachers, it may be wise to consider how mentoring dialogues could be 'stage managed' or structured in ways that ensure that the students will use them to try out their new pedagogic understandings in the public language in which these understandings are most powerfully represented.

Evidence gathered in classrooms which relates to previously agreed discussion points should assist the students' selection of what is relevant to their current learning needs. At the same time it should provide a relatively unthreatening focus around which mentors might open up discussions that are designed to encourage as joint action the exploration of particular elements of pedagogy. In addition, as students become more experienced they would gain more responsibility for selecting the focus of discussions and would take more control of the conversations, to the extent that they would be in a position to challenge the interpretations of others and assert their own substantiated claims to knowledge.

There may be an impact on teacher mentors from using this set of processes. The data from our three year project showed that teacher-mentors were not opening up their own practice to scrutiny in ways that would allow the exploration of puzzles or dilemmas in that practice (Edwards and Collison, in press). Like the student teachers they wanted to see teaching as a private act. Data gathered by both teachers and student teachers from similar activities may enable teachers to talk more openly about their own practices with students to the mutual benefit of both.

Three important features of these proposed practices require some emphasis. First is the compatibility of a growing confidence in the use of the language in which the knowledge base of teaching is carried with both neo-Vygotskian theories of teaching and learning and with understandings of how informed reflective practice or action research might develop beginning teachers' thinking about practices and their contexts. Related to this is the move in and out of the public and private arenas that is a feature of both frameworks. Second, the ambiguity in the role of the teacher-mentor is recognized. As Wilkin (1995) has argued, mentors are teachers of student teachers. They are, therefore, not to be immediately equated with critical friends, though that relationship might develop over time. Consequently conversations have to be knowingly constructed to manage the learning of beginning teachers in ways that enable the beginners to employ the language they need to use. Finally, the strategies outlined should meet a concern evident in the detailed work of Korthagen and Wubbels (1995) who note that training programmes that are largely based on loosely structured processes of personal reflection may be enormously beneficial to some students but cause considerable problems for those students who like to see a clear structure to their courses.

Concluding Points

It would be naive to suggest that what is being proposed here would be easily achieved (Edwards and Collison, 1995). What is not naive is, however, to recognize that despite intentions in the UK to create ITT programmes that produce reflective practitioners — and here one would include the programme that formed the centre of the study of mentoring discussed here — there is a dearth of evidence to indicate that the programmes have been successful.

What is being proposed in this chapter is a way of bringing reflection on practice into the mainstream of programme planning by connecting it closely to a framework for understanding how students may best learn in the settings in which they are placed during their training. One way of ensuring that it remains central might be the development of students as self-aware learners through the use of learning contracts, a practice that is already becoming widespread. However, the most important feature of the processes outlined would be the benefit to be accrued by schools as teachers make the most of the professional development opportunities to be experienced while being a mentor. If schools were to seize the opportunity and teachers were to be encouraged into this method of mentoring we might see a form of ITT which would transform the whole of the education system.

References

BENNETT, N. and CARRE, C. (1993) (eds) *Learning to Teach*, London, Routledge.

BRUNER, J. (1983) *Child's Talk*, Oxford, Oxford University Press.

COLLISON, J. (1994) 'The impact of primary school practices on student experiences of mentoring', BERA Annual Conference, Oxford, September.

COLLISON, J. and EDWARDS, A. (1994) 'How teachers support student learning', in REID, I., CONSTABLE, H. and GRIFFITHS, R. (eds) *Teacher Education Reform: Current Evidence*, London, Paul Chapman, pp. 131–6.

DFE (1993) *The Initial Training of Primary School Teachers*, Circular 14/93 (England), London, DfE.

EDWARDS, A. (1994) 'The curricular applications of classroom groups', in KUTNICK, P. and ROGERS, C. (eds) *Groups in Schools*, London, Cassell, pp. 177–94.

EDWARDS, A. (in press) 'Teacher education: Partnerships in pedagogy?' *Teaching and Teacher Education*.

EDWARDS, A. and BRUNTON, D. (1993) 'Supporting reflection in teachers' learning', in CALDERHEAD, J. and GATES, P. (eds) *Conceptualising Reflection in Teacher Development*, London, Falmer Press, pp. 154–66.

EDWARDS, A. and COLLISON, J. (1995) 'Partnerships in school-based teacher training: A new vision?' in MCBRIDE, R. (ed.) *Teacher Education Policy: Some Issues Arising from Research and Practice*, London, Falmer Press.

EDWARDS, A. and COLLISON, J. (in press) 'What do teacher mentors tell student teachers about pupil learning in infant schools?' *Teachers and Teaching: Theory and Practice*.

ELBAZ, F. (1990) 'Knowledge and discourse: The evolution of research on teacher thinking', in DAY, C., POPE, M. and DENICOLO, P. (eds) *Insight into Teachers' Thinking and Practice*, London, Falmer Press, pp. 15–42.

ERAUT, M. (1995) 'Schön shock: A case for reframing reflection-in-action?' *Teachers and Teaching: Theory and Practice*, **1** (1), pp. 9–22.

GRIFFITHS, M. and TANN, S. (1992) 'Using reflective practice to link personal and public theories', *Journal of Education for Teaching*, **18** (1), pp. 69–84.

KORTHAGEN, F. and WUBBELS, T. (1995) 'Characteristics of reflective practice: Towards an operationalisation of the concept of reflection', *Teachers and Teaching: Theory and Practice*, **1** (1), pp. 51–72.

MACKINNON, A. and ERICKSON, G. (1992) 'The roles of reflective practice and foundational disciplines', in RUSSELL, T. and MUNBY, H. (eds) *Teachers and Teaching: From Classroom to Reflection*, London, Falmer Press, pp. 192–210.

MAGUIRE, M. (1995) 'Dilemmas in teaching: The tutor's perspective', *Teachers and Teaching: Theory and Practice*, **1** (1), pp. 119–31.

MCINTYRE, D. (1993) 'Theory, theorizing and reflection in initial teacher education', in CALDERHEAD, J. and GATES, P. (eds) *Conceptualising Reflection in Teacher Development*, London, Falmer Press, pp. 39–52.

MCINTYRE, D. (1995) 'Initial teacher education and the work of teachers', in RUDDOCK, J. (ed.) *An Education that Empowers: A Collection of Essays in Memory of Lawrence Stenhouse*, Clevedon, Multilingual Matters/ BERA, pp. 29–43.

MILLER, J. (1990) *Creating Spaces and Finding Voices*, New York, SUNY.

PRISK, T. (1987) 'Letting them get on with it: A study of unsupervised talk in an infant school', in Pollard, A. (ed.) *Children and Their Primary Schools*, London, Falmer Press, pp. 88–102.

Ruddock, J. (1992) 'Practitioner research and programmes of initial teacher education', in Russell, T. and Munby, H. (eds) *Teachers and Teaching: From Classroom to Reflection*, London, Falmer Press, pp. 156–70.

Schön, D. (1983) *The Reflective Practitioner*, New York, Basic Books.

Schön, D. (1987) *Evaluating the Reflective Practitioner*, San Francisco, Jossey Bass.

Shotter, J. (1993) *Cultural Politics of Everyday Life*, Milton Keynes, Open University Press.

Shulman, L.S. (1987) 'Knowledge and the foundations of the New Reform', *Harvard Educational Review*, **57**, pp. 1–21.

Thapp, R. and Gallimore, R. (1988) *Rousing Minds to Life*, Cambridge Mass, Cambridge University Press.

Tickle, L. (1993) 'The wish of Odysseus? New teachers' receptiveness to mentoring', in McIntyre, D., Hagger, H. and Wilkin, M. (eds) *Mentoring*, London, Routledge, pp. 190–205.

Vygotsky, L.S. (1962) *Thought and Language*, Cambridge Mass, MIT Press.

Vygotsky, L.S. (1978) *Mind in Society*, Cambridge Mass, Cambridge University Press.

Wertsch, J.V. (1985) (ed.) *Culture, Communication and Cognition*, Cambridge Mass, Cambridge University Press.

Whitehead, J. (1985) 'An analysis of an individual's educational development: The basis for personally oriented action research', in Shipman, M. (ed.) *Educational Research: Principles Policies and Practices*, London, Falmer Press, pp. 97–108.

Wilkin, M. (1995) 'The context of mentoring'. SRHE Mentoring Network Seminar London, SRHE.

10   Discussion of a Pilot Project to Focus the Thinking of BEd (Primary) Students

Sue Cox

The Foundation Teaching Studies (FTS) course at The Nottingham Trent University developed out of alternative ways of thinking about the way in which students might learn how to teach. It is a traditional approach which takes students through a number of discrete course components in the university interspersed with blocks of teaching practice in schools where they are expected to apply the knowledge they have gained. This raises issues about what is important to know and in what order — decisions which are made by the university and which may not relate to the perceived needs of the student. As the students see it, this widens the gap between what goes on in the university and what they are centrally concerned with, which is the activity of teaching children in schools. In broad terms this may be seen by the students as a gap between 'theory' and 'practice'.

It seemed important to take on board the fact that the students' current experience of school on entering the course is already substantial and they already have ideas about what is involved in teaching. In planning the FTS course the intention was to find ways of bridging the 'theory–practice' divide and to find more student-centred ways of designing the course, acknowledging that individuals' existing knowledge could be a valid starting point. A better perception of relevance would be achieved by structuring the course around students' activities in school. Whilst this was not to be a content free, entirely process-based course, it was intended that students would be able to progressively make sense of their teaching experiences in school, starting, in a sense, with the undifferentiated 'whole', rather than attempting to construct a 'whole' from a series of predetermined elements. A series of 'focal points' would provide the content, drawing the students' attention to aspects of children's learning which mirror the model of learning underpinning the course. The principles and processes of the course were to be consistent with those of action research, thus laying the foundations for the action research project which students would undertake in their fourth year, and establishing the critically reflective approaches to learning which were deemed appropriate both for the BEd course and for the students' future professional careers.

The FTS course is based not only on a realignment of the relationship between theory and practice, but on the kind of redefinition of it that may be familiar to action researchers. The traditional conception of the relationship between theory and practice is essentially an instrumental one. Theory generates (usually through positivistic empirical research) generalizable knowledge claims that can then be applied to practice. The plausibility of this view has its origins in the idea that hypotheses are postulated and experimental processes are conducted outside any evaluative framework and hence theories are descriptions of objective reality. Whilst this position has been abandoned by all but the most thorough positivists, there remains the assumption that the activity of theorists is concerned with describing and explaining the world and is to be distinguished from the making of judgments. The implications of this view are that the activities of theorists are distinct from those of the practitioner in that the latter, being essentially engaged in action, must make decisions about what to do, and deciding to do one thing rather than another necessarily entails making judgments.

The problem of how to come to terms with this divide with respect to educational theorists and educational practitioners has always been a pressing one, since education is a practical business and educational theory is necessarily called upon to have value in practical terms. In the past it has been addressed in a variety of ways. For example, O'Connor (1957) did not admit educational theory as theory, claiming that it could only be a 'courtesy title'. Hirst (1970), on the other hand, argued the case for claiming a parallel between educational theory and scientific theory (as defined by Popper (1972)) giving it a distinctive status as 'practical theory'. This has not, however, effectively overcome the theory/practice dichotomy. An alternative approach is to fully take on board the fact that all theorists are themselves engaged in activity.

As with all action, intentions are logically implicit and hence theory, or rather, the activity of theorizing, cannot be divorced from the evaluative assumptions that shape those intentions. The shift, then, is away from theory as logically prior to practice. Theorizing is another form of practice. Since beliefs, ideas and value positions are necessarily incurred in this and all other forms of practice, theory can be construed in this sense — as the beliefs, ideas and values implicit in practice. This was a line of argument that was pursued by Ryle (1949). He effectively exposed the implications of seeing action as logically dependent on theory. He argued that if theory has logical priority over practice then theorizing, itself a form of practice, is dependent on some prior theory. On the same argument, this theory presupposes a theory and so on. We enter a situation of 'infinite regress'. To avoid this *impasse*, he argues that intelligent actions can be seen to derive their meaningfulness and sense from the principles and rules governing them in practice. We learn how to act within social and cultural contexts which govern the appropriateness of our actions. Such know-how is not logically dependent on propositional knowledge, rather, the latter is embedded within the former. To use Schön's (1983) term, it exists as 'knowing in action'.

Figure 10.1: *The reinterpretation of the relationship between theory and practice*

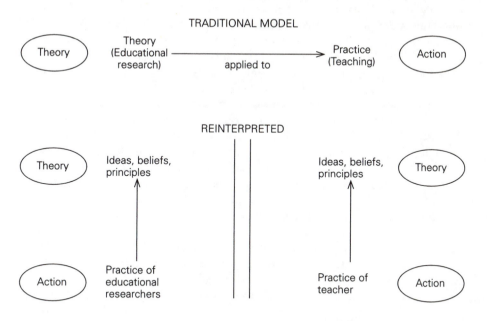

Carr and Kemmis (1986) likewise adopt the view that the activity of educational theorists and the activity of teachers are different spheres of practice, each with their own implicit theory. On these grounds, they argue that it is the theory which is implicit within their own practice, that it is of concern to teachers and that it is misguided to attempt to bridge the gap between the activities of theorists on the one hand and the activities of teachers on the other.

If we persist with a traditional view of theory it leaves unanswered those questions about the values to which it is being instrumentally 'applied'. It cannot provide answers to the questions about what we ought to be doing and what paths of action we should take. It cannot do so since it is deemed to be descriptive and explanatory and thus leaves things as they are. The traditional account of theory and its relationship to practice, then, fails to address crucial issues about the aims and purposes of teachers' actions, unlike the alternative account, which conceives theory in terms of *the* intentions implicit in actions which are carried out within whatever evaluative paradigm gives them meaning and sense. As MacIntyre (1988) claims, 'Forms of social institution, organization and practice are always, to greater or lesser degree, socially embodied theories.'

This reinterpretation of the relationship between theory and practice (illustrated in Figure 10.1) negates the idea that theory must be acquired as a prerequisite of practice; that students apply the theory they acquire in the university to the practice they carry out in school. This is logically untenable on the definition of theory, I have already outlined. Rather, the starting point must be their ability to take part in that area of activity we call teaching, following codes and patterns of behaviour and being able to act appropriately, albeit more or less intuitively. They will have acquired this practical knowledge or 'know how' from being participants in the social processes that embody the values, rules and procedures of this sphere of activity, though they may not be aware of what these are. The aim, then, in terms of the students' learning, is to examine their actions to discover what these implicit norms might be and the nature of the social context in which they are followed. In this way, theory — or understanding of what they are about — is to be generated from within the frame of their own individual experience and practice. Unlike in the instrumental approach, the theory generated is centrally focused on the aims and values that underlie the way things are done within a particular context. It does not purport to give a value free account of an objectified reality. Rather, it acknowledges the contingency of reality.

It could be argued, then, that some of the difficulties faced on a more traditional course, conceived around the idea of 'applied theory' can be attributed to the fact that students are working in school where they are constantly making decisions on how to act, based on judgments made in unique situations where a complex range of factors come into play. Theoretical knowledge, in the instrumental sense, not only cannot give them any help in making those judgments or decisions, since it is supposedly descriptive and explanatory, but being such, it cannot, in its generalized form, help them to identify the features of the specific and highly complex real life context.

On the other hand explication of the theoretical underpinnings of actions by the practitioner — the person who carries out those actions — allows her or him to understand the meanings of those actions and also to change direction in the light of re-evaluation of these ideas and beliefs. Actually attempting to make explicit, through reflection and analysis, the beliefs that are initially implicit, might be a first step towards this, so that the assumptions underlying imitative or intuitive actions are exposed. However, whilst the individual is acting within their own framework of understandings, appraisal of the principles of procedure underlying their practice may result in reaffirmation of existing approaches rather than possible reformulation of them.

Clearly, to address this, pedagogical processes that could encourage the students to become more critically aware of their own framework of reference were required on the FTS course. One way was to provide specific inputs in the university-based elements to help the students understand how their ideas and beliefs are connected to the norms generated within social and cultural frameworks of value. For instance, the way in which they make judgments about children is related to social constructions of race and gender. Another way was to create a variety of settings and relationships through which the student can encounter alternative frameworks of reference and engage critically with them. These structures and processes have developed over time. Figure 10.2 illustrates the model of learning envisaged and Figure 10.3 the

Figure 10.2: A model of students' learning

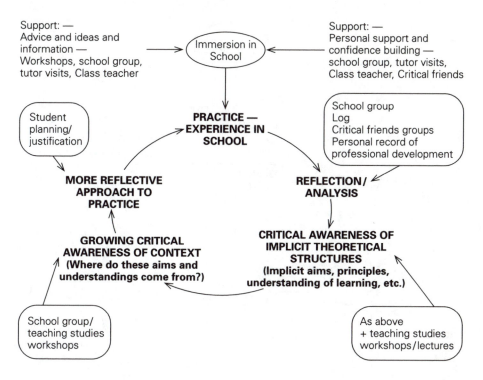

settings and relationships adopted. The school group 'forms the direct link between school and university.' These are small groups made up of three or four critical friends groups. The tutor working with the students visits them in school and so is familiar with the setting in which they are teaching as well as facilitating discussions in 'school group meetings' in the university.

Fundamentally, then, the course embodies a constructivist approach to learning, which acknowledges that individuals make sense of the complex web of cultural practices by constructing their own ideas on the basis of their existing knowledge. This is progressively reconceptualized through further experience of acting upon and within the social context.

Personal Records of Professional Development

The Pilot Research Project

The intellectual challenge of the course lies fundamentally in recognizing that teaching is problematic. On the view of educational practice and theory explained above, the 'problems' that will be encountered by an individual student cannot be defined outside the immediate practical context and what

Figure 10.3: Settings and relationships adopted

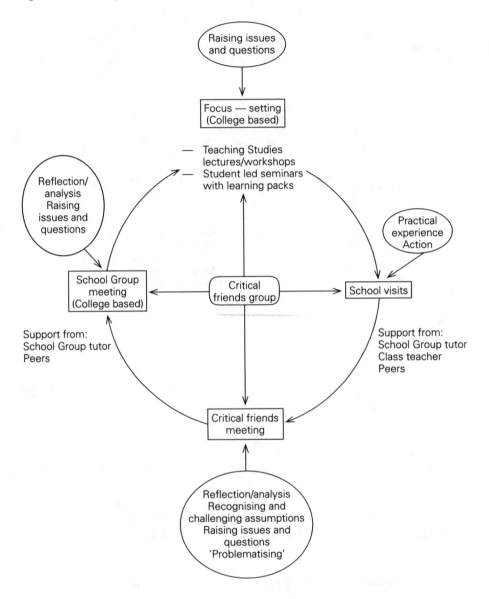

is deemed to be problematic by the student teacher will depend on the way in which they construe the situation in terms of the particular beliefs which they bring to bear upon it.

Many teacher educators are familiar with the situation where a student is happy to report that they have no problems. The student sees this as the achievement of their goal. What is left unexamined, here, is the package of assumptions which was necessary to support the student's interpretation of the situation and their perceived success.

Ongoing evaluations of the FTS course revealed that this, and related issues, were emerging as significant problem areas in relation to the aims of the course. The students appeared to hold expectations that they would be instructed in how to teach and they appeared to have previously acquired attitudes to learning which could not encompass uncertainty and open-endedness. My interpretation of these concerns was that the students tend to hold a technicist view of teaching as knowledge transmission both in terms of what they expect to be doing in school and what they expect to be happening to them in the university. They look to the 'experts' for definitions of the problems they will encounter in how to teach, with the assumption that ready made solutions exist — it is just a matter of the experts passing on their knowledge. This 'deficit' view of teaching and learning does not, of course, encourage students' confidence in their existing 'know-how'.

This, as pointed out previously, is where we must start: with the students' present conceptions and concerns. Clearly there was a need to investigate more closely the processes whereby we might challenge the assumptions underlying these views. At the same time we needed to demonstrate that their existing practical abilities were a valid starting point. We needed to help students to see the value in experiences which were not immediately reconcilable with their expectations — both in relation to their own teaching activities and their experience on the course, and we needed to help them to use these experiences constructively. Recognizing dissonance of this kind — what we could call problematizing their experience — calls into question their actions and responses and, concomitantly, their aims and purposes, as learners and teachers. Fundamentally we needed to raise students' awareness of the importance of 'problematizing' teaching in the development of the quality of their practical judgments. In investigating the processes that might meet these needs we might be discovering whether our aim, to initiate an action research approach to learning to teach, was achievable at this early stage.

The pilot project was carried out as a piece of qualitative research, using action research methodology. It was decided to look at the role of group work which has always been a feature of the course. The use of critical friends groups was given more emphasis with students meeting independently of tutors to reflect on their experiences in school the previous day. In view of the problem areas identified above, the desirability (in terms of the coherence of the course) of giving critical friends group meetings a common direction and purpose, and also the fact that students had repeatedly claimed that they did not know how to be analytical about their practice (though all were able to *describe* it), it was decided to develop some sort of specific means of focusing students' thinking.

Students had already been introduced to a way of recording their learning, identifying in a cyclical way what they were learning, why they saw this to be important and what they needed to give attention to next on a series of pro formas. This was deemed to be supportive to the students in that it enabled them to reflect on their personal concerns and to record the way in which their learning progressed. Given the way in which initial teacher training courses are being encouraged to assess students' performance in the classroom by way of discrete 'competences' (often conceived in the form of lists of performance indicators over which the student has no ownership), this seemed one way of helping students to identify their own developing competence. In order to increase this sense of personal ownership and for students to develop their own profile of their learning, the idea was developed in the form of a 'Personal Record of Professional Development'. This was to take the form of a series of pro formas and were constructed in such a way as might encourage the students to 'problematize' their teaching. On the analysis of educational theory I have presented, it seemed that a possible way forward might be to ask students directly to attempt to articulate the assumptions which underlie the actions which they spontaneously and 'intuitively' carry out. Over and above this articulation of alternative frameworks of 'theory' implicit in alternative courses of action might enable the student to evaluate his or her own actions and his or her own beliefs and ideas in the light of those of others with the possibility of change in both their practice and their way of thinking. See the questions below.

The intention, then, was that the 'Personal Records of Professional Development' would record 'critical incidents' in students' experience which would occasion the kind of reflection we wanted to foster. With the help of critical friends they would examine the tacit understandings which they brought to bear on the situation and identify further courses of action to bring about change and to reconstruct their ideas. This would be an ongoing record which would document the path of an individual's learning during the course.

Students were asked to complete one of the pro formas after each visit into school (in which they were expected to undertake some work with children). On the following day in a meeting with the small group of critical friends (about five students in the group) they could gain a response from at least one of these fellow students to acquire a different interpretation of their own actions and to develop 'alternative perspectives'. The pro formas were introduced and explained to the students by their school group 'tutor'. They were also given an explanation of how to use them in a large, lecture style presentation. A written example was provided for reference.

Some of the Main Issues Arising from the Research

In the first round of evaluation, in which all students were asked to comment on the Critical Incident sheets, some of the students responded positively to the activity. For example:

> I find that the Critical Incident sheets are helpful in that they encourage us to evaluate our time in school in a way that we probably wouldn't do otherwise.

CRITICAL INCIDENT SHEET

Brief description of the context

Brief description of what was happening

What was my part in this?

What do my actions tell me about my way of seeing things?

Own comments

Critical friends' comments

Any other way of seeing things?	How could I have done things differently?
Own comments	Own comments
Critical friends' comments	Critical friends' comments

What problem(s) or issue(s) am I dealing with?

Own comments

Critical friends' comments

What next?

> The questions are broken down well to give you ideas to evaluate which is better than just writing an overall evaluation.

> Excellent idea to force you to evaluate things you might otherwise overlook.

Since not all students saw the process to be beneficial the negative responses need to be analysed. These were mostly related to:

1. motivation;
2. difficulty in interpreting the activity.

These are by no means mutually exclusive categories.

Motivation

With regard to motivation some students found the Critical Incident sheets rather daunting. It may be that it was the form of the sheet itself rather than the process of analysis that was off-putting, but there were clearly difficulties in understanding the sheet (see above). There were practical difficulties in that they took too much time to fill in. Some students said that they preferred to use the critical friends meetings for open discussion about their day in school and did not perceive the sheets as a useful vehicle for discussion. They constrained, rather than facilitated, the discussion. It was already becoming apparent that the process of filling in the sheets did not suit all students — that there was a need to take fuller account of individual learning styles. For example, some students needed to write in order to clarify their thought, others preferred to talk. There was concern that there was no assessment of this particular aspect of their work — this was cited as an explanation of lack of motivation.

Difficulty in interpreting the activity

A common response was that the sheets were 'repetitive' and required them to say the same sort of things over and over again. This seemed to indicate difficulties in interpreting what they were asked to record in the different sections of the sheet. For example:

> The questions are somewhat badly phrased and go over what's been said in the previous section.

> I find them rather repetitive and seeming to keep asking the same questions.

Another major issue was the way in which students perceived the purpose of the sheets. It was clear that in many cases the students felt they were recording 'problems' that were immediately apparent and that they had to 'deal' with rather than problematizing their practice. For example:

The Critical Incident sheets are helpful if you have problems to fill them with. Otherwise it seems pointless filling one in *when nothing actually happened.* (My emphasis)

If you have no problems then it becomes quite hard to fill them out.

I feel that I am 'waiting' for an incident to happen so that I can fill in a sheet. I am sure that this is not the right thing to do.

It is nice to receive both support and ideas on how to deal with such an incident if it arises.

This was also evident in a variety of other forms of response that commonly occurred. For instance, a number of students felt that their critical friends could not help them, as they were no more experienced than they were. For example, a typical response was:

Where many of us have experienced the same problems we have been unable to help each other. It is difficult to critically analyse someone else in the group when we have had so little experience.

In some cases the perception seems to be that the critical friend's role is to evaluate the student's own response or action, rather than offering an alternative perspective, as is illustrated by this student's response (extract from Critical Incidents sheet):

I agree, these ideas are good.

Or the role is to suggest to the student the 'solution' to a 'problem' or the 'correction' of a 'mistake'. For example: (extract from Critical Incidents sheet)

You should have encouraged the children to do their set work to ensure that they completed the task set.

In other cases the perception is that the critical friend will offer support of a general kind:

The teacher should have taken responsibility for this child; it shouldn't have been left up to you at this stage.

What all the above examples seem to indicate is that students are looking for the 'right' course of action. They tend to confirm the view that learning is perceived by the students on a deficit model, in that they see their own knowledge and performance to be inadequate as measured against some established norm, embodied in the knowledge and performance of the 'experts'. This is revealed, for instance, in those cases where their critical friends are regarded as not having the level of knowledge required to help them 'get it right' and where on those occasions when the critical friends did try to offer support it was not in the form of an interpretation of the student's actions from an alternative point of view, but in the form of an evaluation of them against some apparently unidentified criterion of what is a good or right action. Where a 'problem' is identified, there are implicit beliefs about the inadequacy of their practice against some perception of the right way of doing things, but again the assumptions underlying the 'correct' solution are not necessarily either acknowledged or questioned.

It seems to me that there are two ways of looking at this related to different ways in which a 'problem' is recognized. On the one hand it may be identified through the mismatch between the outcome of a student's action and her or his intention (the intended outcome). Whilst this may well prompt the student to articulate her or his original intention there is no necessary implication that she or he will address the assumptions underlying this intention. On the other hand, the outcome matches the student's intention, but in the eventuality gives them reason to question their action by revealing features of it in practice which in some way conflict with their broader conception of what they are about. Again, this is likely to be a conception based in a taken-for-granted normative view which in itself remains unacknowledged and not subjected to any critical scrutiny.

Now it is possible that this process may involve consideration of a variety of alternative courses of action and decisions about what one ought to do in the circumstances. To this extent it implicitly identifies aims or principles of procedure. Nevertheless these *remain tacit* unless, of course, the students are spontaneously questioning each other about the kinds of beliefs and ideas which inform their judgments. It has to be allowed that this may, in fact, be going on. It may be that the student is in fact engaging in the processes of analysis that are intended by the 'critical incident' sheets, incidentally to the process of solving a specific problem. This remains to be investigated. But, to the extent that the process is seen as solving specific problems and finding the 'right' solution, the possibility persists that tacit understandings remain unarticulated. Furthermore, to the extent that alternatives that might be offered by a critical friend are disregarded, the student is bound by the limitations of her or his own perspective (unless the student is spontaneously questioning her or his own framework of reference), as in the following example:

> One can see where one went wrong after the event and so usually
> you have sorted it out for yourself and do not need to discuss it.

Fundamentally, an approach to analysis in terms of problems and solutions leaves those aspects of their practice which students did not see as throwing up problems — and the matrix of beliefs in which they are embedded — unexamined and unquestioned. That there is this tendency for students to perceive the analysis in terms of 'problems' to 'deal with' is further underscored by the frequency with which they refer to the inclination to invent a problem when one was not immediately apparent. A typical response in this case was:

I often feel you're having to make up incidents to fill in the form.

Once again, further evidence suggested that what goes 'wrong' is something that requires 'handling' and putting right rather than being something which prompts reflection on the questions: 'In terms of what beliefs and ideas did I interpret this as going wrong?' or 'Within what framework of beliefs and ideas would my action be inappropriate?' and 'Might there be other ways of perceiving and conceiving things through which I could offer a different interpretation?'

As an extension to the evaluation, a group of student representatives canvassed the opinions of the group as a whole. They supported the principle of encouraging students to 'examine their role as a teacher and their attitudes to situations arising, but their suggested format for analysis was:

What was happening.

How I coped.

Comments (including those of critical friends).

I would suggest that the use of the word 'cope' is particularly significant here in relation to the way in which the students might conceive the task of teaching on the one hand and analyse their teaching on the other, within an unexamined existing framework of assumptions. This same group reported the difficulties students were having in interpreting the questions and cited examples of students manipulating the data in order to complete them. As they pointed out 'at best this was considered to be unnecessarily time consuming and at worst unproductive.'

It seemed that the Critical Incident sheets were not fulfilling the aim of providing a structure that would enable students to focus on the implicit theoretical constructs underlying their practice. However, analysis of the students' responses to and use of the sheets did reveal more clearly the students' conception of teaching in terms of getting it right. I would also suggest that it does indicate that at least some students do, indeed, see learning as a linear process, that they see themselves as floundering in the dark not knowing what to do, and that an answer can be provided. As discussed earlier, it was these concerns and conceptions that I had originally hoped to address through the process of analysing critical incidents. The means I had chosen were clearly inadequate to the task of challenging these and building an understanding of the dynamic nature of the process of learning. It did not, as it stood, enable

the students to conceive their learning in terms of the development of strategies for making sense of teaching which could be modified in a whole variety
of ways in the context of identifying and refining purposes in the critical
company of others. The priority was to evaluate performance narrowly rather
than understand the range of meanings of actual and possible courses of
action. I have argued that through doing the latter, the process of evaluation
can go much further. But this was not, at present, being achieved.

At this point it seemed that the notion of 'critical incident' was itself
misleading for the students. Although it is useful to the extent that it can refer
to an event which marks some sort of dissonance that might trigger useful
kinds of reflection about what one was doing and what one's actions might
mean, it seemed rather to be interpreted by students as an occasion when
something went wrong. It seemed appropriate, at this stage, to redesign the
pro forma: (a) to simplify it (to respond to students' concerns); (b) to encourage students to consider aspects of their practice over and above the 'problems' identified in terms of their preconceptions — to, help students to see that
what they take for granted, and which, does not throw up any further 'problems'. The taken-for-granted only remains unproblematic when it is perceived
as the only way of looking at things. Once an alternative perspective is introduced, the student must accede that there are choices which have to be made
about how to proceed which entail evaluation of different frameworks of
belief and understanding. To this end the language of 'critical incidents' was
replaced with the language of narrative description, see below.

Outcomes in terms of the evaluation of the redesigned format are incon-
clusive. Students were specifically asked to comment on what they thought of
the Classroom Practice Analysis sheets in principle and most of them could see
their value. For example:

> The principle of analysing situations and classroom issues is valuable.

> I feel that the Classroom Analysis sheets are an essential core to pro-
> fessional learning and development. Without this process of enquiry
> and observation within the classroom, many important issues would
> not be raised or evaluated or reflected upon. They act as a good
> resource for issues which may arise in the future. I feel the principle
> of having a means of examining one's own practice is important.

Though there were no entirely negative views 'in principle' in the sample of
thirty-six students analysed, one student commented:

> I don't think much of them in principle. I can't actually see their point.
> They just seem to waste valuable time . . . I always seem to be filling
> them in because I've been told to and getting nothing out of it at the
> time or in the future. (The student did go on to say that discussion
> with fellow students was 'a good idea'.)

Personal Record of Professional Development

CLASSROOM PRACTICE — ANALYSIS SHEET   DATE _____

STORY (Include, for example, the context; what was going on. Ensure that you include your part in the situation)

Write this section independently

IMPLICIT THEORY (What do your actions/responses tell you about your way of seeing things, your ideas, beliefs, theories?)

Write this section with critical friend

ALTERNATIVE PERSPECTIVES

Other possible ways of *seeing* things: different ideas; beliefs; theories.	Other possible ways of *doing* things (Alternative courses of action; ways of responding)

Write this section with a critical friend.
The theory and practice sections should be related but either can be filled in first

WHAT NEXT?

What will you do to further develop your practice in relation to the issues/problematic areas identified?

Write this section independently

FURTHER ANALYSIS / COMMENTARY

Fill this in as appropriate. Continue on extra sheets as required

The students' comments on the Classroom Practice Analysis sheets revealed a similar range of difficulties in the second round of evaluations as the first, although there was some general agreement that this format was more user-friendly than the Critical Incident sheet format. It has to be remembered that students had already interpreted the analysis in terms of 'problems' or 'mistakes' and this attitude persisted, as expressed in such comments as:

There isn't always something to write about.

They are a bind. Sometimes as I feel I have already gone about tackling an incident and ensuring the same mistake will not happen again.

I often have trouble identifying problems.

This view was also revealed in comments on 'the principle'. For example:

Essentially they seem useful in regards to isolating a problem incident and coming up with a solution.

They seem a good idea, but most people seem aware of the mistakes they've made.

This points again to the students' tendency to be preoccupied with 'getting things right' in the classroom. The introduction of the 'Personal Record of Professional Development' may not be enough in itself (even within the structured framework of critical friends groups) to help the students to see that 'performance' is necessarily appraised against criteria — whether implicit or explicit — that are embedded in a matrix of beliefs, ideas and theories. It did not, it seemed, enable them to engage in reflection on this level.

The outcomes of the project thus far indicated to me that I needed to further examine my own assumptions about the efficacy and value of the structure of the analysis as a way of focusing students' thinking. I presented the work at a round table session at the Collaborative Action Research Network (CARN) conference (Birmingham, 1994) inviting participants to offer alternative interpretations of what I was doing and alternative courses of action with their concomitant frameworks of theory. It was brought to my attention that the process I had introduced essentially had a cognitive focus — being directed at students' thinking — and perhaps overlooked the affective dimension. On reflection it is clear to me that there was a powerful emotional response in the students' evaluations of the scheme of which I was aware, but to which I was not giving due attention. There were clearly ways in which the sheets were seen by some students as threatening, demanding and difficult to interpret, which were perceptions that should not only have alerted me to the students' implicit conceptions of teaching and learning but to the feelings of anxiety, frustration and possibly hostility in some cases. Whilst some measure of affective discomfort is arguably part of any learning situation, one has to question the extent to which this would affect motivation and thereby limit the possibilities of enhancing learning.

On another level, the sheets themselves required the students to describe what was happening; what their part was in this. Though this leaves the way open to recording their feelings they are then directed to interpret their *actions*. Reflections on what they felt, likewise, can reveal underlying theoretical constructs. This presumption necessarily rests on an account of emotional experience in which thought and feeling are interrelated; that the emotions which a person experiences can be seen to be related to a cognitive appraisal of the objects of the emotion or in other words that the emotions have a 'cognitive core' (Peters in Dearden, Hirst and Peters, 1972). It is because a person thinks about things in certain ways that they feel the way they do. Thus, for example, the student who felt frustrated by the children who did not carry out her instructions to use their imaginations was able to both acknowledge her belief that children ought to do what she asked them and her theory that children could be instructed to use their imaginations. On reflection she considered alternative perspectives that were offered, that it might not be a reasonable expectation for children to carry out her instructions. Her theory that 'imagination' was some kind of pre-existing faculty that children could tap into could be challenged by the view that prerequisites of imagination are experience and understanding and more than the simple instruction was needed to enable the children to draw upon them. Perhaps, also, there is some level of contradiction in expecting children to respond to her instructions on the one hand (the message here being that someone else is in control) and to be imaginative on the other (the message here being that they are to think for themselves). I would suggest that these insights, generated in response to the intensity of her feeling, might not have been gained from a dispassionate account of her actions.

There are further questions to be addressed about the relationship *between* feeling and action. To the extent that the motivation to act in certain ways can be based in the emotional domain, and that ways of feeling are themselves socially constructed, our actions can be interpreted in terms of implicit ways of feeling which can be questioned and challenged. Once again, however, there would be a 'theoretical' or cognitive component, since the idea of development of, or changes in, ways of feeling, depends on a view of emotions as dependent upon ways of 'seeing' or understanding. The 'story' format of the redesigned sheets can be used to encourage students to record their feelings fully, and I acknowledge that students should be encouraged to do so.

At this point in the research a range of further issues that could be addressed were becoming clarified:

- The practical issues of time allocation, etc.
- The possible need for more extensive tutor modelling of the process.

- The messages carried by the use of the pro forma itself and the need to be more responsive to individual learning styles. (For example, some students need to write in order to clarify their thoughts; others prefer to talk.)

Taking account of all of these, I decided to use the structure of the sheets as a way of focusing the students' thinking following occasions where I had observed some teaching practice and was able to discuss this with the student. Students tend to see interactions with their tutors, following observation, as an opportunity to gain some feedback on their performance along the lines of 'How did I do?' Often, the student experiences anxiety, feeling that there are inevitably shortcomings in their practice, measured against the criteria, both implicit and explicit, that the tutor will apply in expressing a judgment. Once again the student is concerned to have 'got it right' in terms of a model of practice which the 'expert' holds. Given the outcomes of the use of the Critical Incident/Classroom Practice Analysis sheets, I decided that an alternative approach would be to talk through the questions on the sheet with the student, acting as the critical friend myself. In this way the student would have the analytical process modelled. It would also take account of the time taken and the difficulties experienced by some students in writing things down. The process of feeding back on the students' teaching would shift from the *judgmental*, whereby the student arrives at an understanding of what constitutes so-called good practice (the key to which is held by the tutor) to the *analytical*, whereby implicit criteria (the theoretical ideas contained within an evaluative statement) are identified by the student and the consideration of alternatives is shared by the student and tutor — once again entailing a shift in emphasis from evaluation, in terms of measurement against some pre-established norm, to interpretation and subsequently to an enriched form of evaluation. In the context of the tutorial, the shift is quite significant in that the student's expectation of the tutor, in line with their perception that the tutor must have the answer, is that the tutor will be confirming or criticizing their performance.

This stage of the research was limited to three of the students on the teaching practice following the FTS course, but there were some interesting outcomes that influenced my thinking about the nature of the tutor–student interaction and the value of the process of classroom analysis I was trialling. One student, who in spite of my best efforts to facilitate self-assessment rather than to pass judgment, had found it very difficult to articulate her thinking about what had been going on and expressed anxiety and lack of confidence, clearly found herself to be in control when we approached her teaching analytically following the classroom analysis process. She was able to identify the assumptions implicit in how she had set up the activity for the children, of which she had hitherto been unaware. She became quite animated and expressive and the indications were that she had been pleased by what she had been able to do. Her comment, 'I wouldn't have come up with all this if I'd had to think of it on my own,' gives some indication of the gains she had made through working with a tutor as critical friend in this way. Another student had considerable ability to talk about her own teaching. Once again, the sense of being in control, the student-centred focus of the analysis, enabled her to elaborate extensively on her ideas, and the classroom analysis process enabled her to clarify her own theories to a level that was quite remarkable in my own experience of second year students. The consideration of alternative perspectives led her in this case to confirm the approach she had taken. By working in partnership with myself, acting as critical friend, however, she took the step of identifying alternatives, rather than simply justifying her own position, which encouraged her to accept that there may be other strategies justifiable within a different framework of reference. Her comment, at the end, that 'I had disliked the Critical Incident sheets as I was so set in my ways I just couldn't think,' seemed to underscore the value of the role of the tutor in acting as a 'critical friend' in this particular way and reaffirms the need for some kind of process which frees the students' thinking. Her further comments, that, 'It's very difficult to act differently to the teacher . . . and that the children dominate how you can act, they do limit you,' demonstrate the contextual constraints which operate on the students' practice, if not their thinking, and point to the need to allow students to step outside these and to explore possibilities.

Whilst I, myself, had always thought that I had previously conducted these kind of tutorials in an empowering way, I found that, following this process, which required me to maintain a neutral, analytical stance rather than a judgmental one, I felt quite different. Throughout the process it was evident that the student was expecting, and wanting, affirmation or evaluative comment of some kind. She was reading implicit criticism into my suggestion that we should consider alternatives, when none was intended. Clearly, for psychological reasons, it is important to ensure that affirmation is offered, appropriately, but it is important that we should resist the tendency to see this as the central purpose of the tutorial, but rather, perhaps, see it as a means of facilitating the process of analysis.

This research has indicated to me how crucial it is, in the process of teacher education, to help students shift their conception of learning to teach from a technical process of solving problems to a reflective process of problematizing action. The research has highlighted the importance of this distinction, which, I would argue, is obscured by the traditional expectations of the students, and probably tutors, about the nature of the process of learning to teach. What needs to be further examined is the framework of beliefs in which these expectations, and the kinds of activities and interactions to which they give rise, are embedded.

The project suggests that it is very difficult to 'disembed' the conception and identification of the problematic from an assumed matrix of beliefs. The challenge lies in finding effective ways of helping students to make explicit the taken-for-granted frameworks, to engage with alternatives and to re-evaluate actions. I would argue that if awareness of such possible alternative frameworks can genuinely be raised then a move has been made towards the development of a more critically reflective position on the part of the student, from which any decision about action is potentially problematic. It would be possible for the student to see that having got it right is dependent on the terms in which that judgment is made — on the matrix of beliefs and understandings which generate the criteria of judgment. The significance of the alternative frameworks of reference which the students can offer each other had not been fully realized through the particular form of the process of analysis I initially introduced. However, the role of the tutor in helping to initiate the students into this form of classroom analysis could be further investigated as this model, on the basis of this research, does, I suggest, have some potential for shifting students' conception of learning to teach towards a reflective process of problematizing action.

References

CARR, W. and KEMMIS, S. (1986) *Becoming Critical: Education, Knowledge and Action Research*, London, Falmer Press.

DEARDEN, R., HIRST, P. and PETERS, R. (1972) *Education and the Development of Reason*, London, Routledge.

HIRST, P.H. (1970) in TIBBLE, J.W. (ed.) *The Study of Education*, London, Routledge and Kegan Paul.

MacINTYRE, A. (1988) *Whose Justice? Which Rationality?* London, Duckworth.

O'CONNOR, D.J. (1957) *An Introduction to the Philosophy of Education*, London, Routledge and Kegan Paul.

POPPER, K. (1972) *Objective Knowledge: An Evolutionary Approach*, Oxford, Clarendon Press.

RYLE, G. (1949) *The Concept of Mind*, London, Hutchinson.

SCHÖN, D. (1983) *The Reflective Practitioner*, New York, Temple Smith.

11 Children's Perspective on Underachievement

David Wilcockson

Introduction

This paper is part of a much larger project undertaken between 1979 and 1991. It grew out of an organizational and administrative need to establish computer records of the children's educational, medical and social experiences. In 1976 this database began to be used to provide qualitative data in order to make sense of the teacher's empirical observations concerning the level of pupil achievement within the middle school of which I was deputy headteacher. Over this period, I constructed a definition of underachievement based on the difference between potential (as measured on an IQ test) and achievement (as measured on a reading test) (Satterly, 1985). This measure confirmed that the school was indeed receiving an unrepresentative range of achievement when compared to the national population at 9+. A further analysis confirmed that whilst the school population over this period was not significantly different from the national population in terms of IQ, it was one standard deviation below both the national and the wider local population in terms of achievement. This was profoundly disturbing as it had been previously assumed that the whole local population was similar and related to the deprived socio-economic group from which it drew (Bernstein, 1971).

A further study of the children's records revealed a strongly positive correlation between poor educational, social, personal and medical histories and underachievement and by applying the Bristol Social Adjustment Guide (Stott, 1976), I was able to establish a link between behaviour and underachievement. In order to clarify my ideas concerning the coding of the variables I decided to review current thinking in the field of underachievement, keeping in mind the intention of comparing the perspectives of teachers, parents and children at a later stage in the research. The research took place in an English middle school, which relates to grades 4–7 in the American system.

Literature Review

There is evidence to suggest that influence of the home may disadvantage children in terms of socialization (Elton, 1989) and that antisocial behaviour, aggression and underachievement can be directly related to the quality of the child's home experiences (Coleman, 1993). Certainly the attitudes of the parents towards schooling can be crucial in encouraging the child's linguistic development, stable attitudes and the desire for academic achievement (Mortimer, 1988).

Spielburger and Gaudry (1974) indicate that a multiplicity of emotional factors can cause underachievement and that poor classroom behaviour may be a result of high levels of anxiety endured by the child (Stott, 1976) both at home and at school. The children's personalities influence the way in which they cope with the threatening aspects of their lives (Thorndike and Hagen, 1969) but high levels of anxiety are linked with poor performance.

Some children may have cultural or bilingual problems that can impede communication between the parent and school (Townsend, 1971) and problems of culture, religion and morality, if not approached positively, can interfere with socialization. This may be a source of anxiety, frustration or violence when children are unable to adjust to different cultural pressures and personal needs (Swann, 1985).

Long-term medical problems (Bullock, 1975) and undiagnosed specific learning difficulties may lead to a cycle of constant failure. This has serious implications for the self-image of the child and poor behaviour may result as a consequence of increasing frustration (Steed and Lawrence, 1988).

There is strong evidence, however to suggest that the way in which the school is organized, the positive relationships of the teachers and unthreatening classroom environment can be powerful forces for encouraging achievement (Mortimer, 1988).

Research Aims

In the light of the earlier work described in the introduction and the subsequent literature review, the aims of the project came to be formulated as follows:

1. To develop a better understanding of underachievement within the school through the perspectives of teachers, children and parents; by sampling, comparing and evaluating their views and experiences.
2. To implement a programme of action research in order to engage teachers, children and parents in a process of change through:
   a. involving all school staff in discussion of the purposes and progress of the project in order to raise awareness and to engender a feeling of 'corporate ownership' in the research;
   b. identifying related teacher inservice training needs and delivering a programme in response to them;
   c. the *processes* of eliciting the perspectives of teachers, children and parents and respecting their current positions;

d. the development of management systems to encourage communication and to facilitate change.

Children's Perspectives on Underachievement

Seeing the school from the pupil's point of view is important for heads and teachers. Knowing what pupils see as positive helps them improve the atmosphere . . . This is a valuable source of management information. (Elton, 1989)

Method

This section describes the method adopted. In summary, all the pupils from a number of classes were involved in a drama project, where they worked in small groups on identifying and expressing through role play what they considered encouraging or discouraging to the learning process. The collective views of each team were recorded and these provided the data on the children's perspectives. The classes were mixed ability and mixed sex (see table 11.1).

The fieldwork took the form of six one-hour drama periods over weekly intervals for the half-term, Autumn 1987. In week six the children were taken to the classroom and after promising to guarantee that they would not miss their drama lesson they settled into their working groups. They were asked to keep in mind the work accomplished by themselves and others during the previous four weeks and to appoint a recorder. Each group was encouraged to record *all* comments.

Task for Discussion:

'What Discourages or Encourages my Learning?'

Organization

Teacher led   Class Discussion
Pupil led   Group Discussion
               Recording
               Review to class by groups

In the final session each child was asked to write his or her own list of answers to the question. One member of each group was designated to act as recorder and to collect on one sheet all the points that each group member had written. These were read out by each group leader to the class to initiate discussion. Forty-two sheets constituted the data for the children's perceptions.

I led the team for the introductory lesson to set the parameters within which the children were to work and for the final classroom lesson in order

Table 11.1: Groups participating in the research

CLASS 1

No of groups	Year	Age	Boys	Girls	Total
5	1	9+	13	10	23
5	2	10+	14	10	24
5	3	11	12	12	24
6	4	12+	15	14	29
21			54	46	100

CLASS 2

No of groups	Boys	Girls	Total
5 Year 1	12	11	23
5 Year 2	14	12	26
5 Year 3	13	11	24
6 Year 4	16	12	28
21	55	46	101

Total number of:
Boys-109
Girls-92
Children 201
Groups 42
Staff-7 (in flexible teams of 2 or 3)

to record the research data. The remaining four lessons were led by other members of staff from the teaching teams. The form of the lessons was that within which the children were accustomed to working:

- Introduction of the task
- Teacher-led class discussion
- Group task–discussion–rehearsal
- Performance to class
- Pupil or teacher-led discussion and review

Four open-ended tasks were chosen to represent the four main influences on children's underachievement as identified from the literature review — The Home, The School, The Child and The Teacher.

Week 1. My worst experience.
Week 2. How I would change the school.
Week 3. Getting my own way.
Week 4. Stopping out late.
Week 5. The children were encouraged to choose one of the previous four tasks to rehearse again in light of later discussions and after viewing other groups' performances. Often children also wished to have time to develop their ideas further and to rehearse for

intergroup performances and year group assemblies. They could also choose their own theme if they wished.

Week 6. Classroom-based group recording.

The open-ended tasks were deliberately designed to encourage a variety of interpretations by the children and care was taken by the teacher leading the initial class discussion to act as an impartial chairperson:

> In such cases the teacher is often leading with hints or suggestions as to the range of contributions being sought. Full and genuine discussion will take place when pupils are given more control over the course of the contributions and indeed when pupils begin to comment on each other's contributions. The teacher's skill in relaxing control over the nature and procedure of the discussion is important here. (DES, 1985a)

The question arises whether the scribes were conscientious in including all the opinions of the group or emphasized their own. Probably there was some bias but the reading out of the 'agreed' lists provided an opportunity for group objection and they would have, in any case, through the group work, assimilated some of the views of their group.

An important point is that I observed underachievers actively involved in the whole process, articulate and, in three cases, actually leading groups. This relates well to the research findings of Slavin (1991) who observed the same phenomenon when studying students' responses to team games. I am confident that the process was at least moderately successful in eliciting the views of underachievers and including them in the recorded material. However, the polled responses precluded a comparison between underachievers and other pupils. Nevertheless, the group approach was justified on pedagogic and practical grounds. Further, the group approach, using the strengths of underachievers, pre-empted the individual attitude of underachievers toward noncooperation and disruption to normal classroom practices (Kerry, 1982).

Coding the Responses

My first concern was to establish relevant categories of pupil response. Initially I had intended to immerse myself in the responses and allow the coding to develop from the perceptions of the children. On reflection, however, it seemed to make more sense to use those categories originally identified in the literature review. In the final analysis the choice was something of a compromise with the major categories being those identified in the literature review and the subcategories relating specifically to the pupil responses. It was a concern to me that I might fall into the trap of developing a fieldwork design in order to prove a predetermined research goal:

> If we could be satisfied that classroom observers were interested in providing reasonably dispassionate information about classroom life and the teaching process then we could accommodate the difficulties attendant upon participant observation . . . But this is not the case . . . they use their field work experiences to promote their own research goals. (McNamara, 1980)

It seemed to make sense to decide on a common approach to coding in order to facilitate later comparison.

I made no attempt to dismiss responses which seemed to be saying similar things. I could not be certain that similar perceptions did not in fact refer to different incidents and so it seemed more honest and accurate to code all responses. Where a sentence contained two or more clearly defined ideas they were coded separately, and where a response could not easily be coded I relied upon the child's use of language and the stress laid upon the sentence construction.

The survey enabled an analysis to be made of the changes in attitude and perceived importance of the variables that occurred over the four years of the middle school (American grades 4–7). It did not enable individual children to be identified or for general differences to be analysed, as the data was generated by groups that were mixed sex and were a composite of responses. However, this guarantee of anonymity was designed to encourage the children to respond honestly and openly.

I was interested in comparing these responses with the perception of teachers (especially their lists of teacher behaviour which could influence pupil achievement). This required a common coding frame and so for analysis a set of categories was devised drawing on the various kinds of factors discussed in the previous literature review.

It would have been particularly interesting to learn the views of other children but the group approach conflicted with trying to get individual responses from identified children. It was certainly clear to me that this project should not work only with underachievers, which would have been labelling and probably ineffectual. The work undertaken on the Stott Behavioural Inventory (1976), suggests that the underachieving child may be unwilling to cooperate, unable to clarify ideas and unwilling to parade opinions for public scrutiny (Kerry, 1982). Further, the literature review suggested that academic failure often led to the underachieving child becoming isolated within the normal classroom where s/he was perceived as a liability in learning and behavioural terms by her or his peers (Stott, 1976). This accorded with the comments of teachers within the school.

Often the underachieving child develops negative behaviour patterns to enable her or him to cope with perceived failure and to support her or his self image. In this way s/he is fulfilling the expectations of those in authority and achieving a personal status to bolster her or his lack of academic success. S/he may also become a negative role model for many of her or his peers (Kyriacou, 1986).

The profile of the underachieving child suggests that gaining her or his cooperation may be a difficult undertaking (Kerry, 1982). Further, investigative research based in the classroom is seeking to utilize a set of parameters within which the underachieving child has already failed. The classroom situation has its own rules of behaviour concerned with acceptable movement, noise levels, relationships and discipline; sets of rules that the underachieving child perceives as relating to failure. The teacher may have also developed strategies for dealing with the disruptive underachiever in order to protect the integrity of the other children and therefore relationships within the formal framework of learning may already be strained with the very children with whom we wish to cooperate.

The research then, to create a positive foundation, was visualized as being conducted on 'neutral' ground away from the classroom and designed to use the creative skills of the underachiever — oral dexterity, provocative questioning, responsiveness to open-ended questions and activities and, when motivated, creative, persevering and inventive attitudes (Kerry, 1982). I was also concerned about the effect of innovative classroom research upon the teacher. The close examination of one's professional performance is personally threatening, and change can threaten control and order. It is perfectly reasonable that teachers should be concerned about this.

It seemed sensible to utilize existing organization and working teams of teachers in order to develop positive teaching situations. Further, an oral or dramatic learning situation encouraged the positive aspects of the underachievers' personality as illustrated on the profile. A further set of advantages accrued from the opportunity to base the research in the drama studio. This would enable some of the constricting rules of the classroom to be suspended. The children were encouraged to become far more involved in the setting of targets, development of ideas and of review and assessment. Using the drama studio also allowed for higher levels of noise, permitted greater degrees of movement and encouraged the opportunity for underachievers to match and compare their experiences without risking judgment or censure. Whitaker (1984) lists the value of such a research setting:

i.   . . . it [drama] creates a climate in which pupils can work with security and self confidence;
ii. it facilitates the growth of understanding by offering the optimum opportunity for pupils to talk reflectively with each other;
iii. it promotes a spirit of co-operation and mutual respect.

Team-teaching was already successfully established for drama, and corporate responsibility for the learning environment was well developed. The teachers involved testified to feeling supported in all aspects of discipline, resource production and organization. Team-teaching also enabled teachers to create opportunities to observe and evaluate the influences of individual children and to concentrate their influence on small groups of children where appropriate.

Utilizing existing teaching strategies and learning situations seemed to me to have positive advantages:

1. the children were accustomed to interacting with the teams and had formed a degree of trust in their motives;
2. the children were comfortable with groups of teachers and were not unduly concerned by changes in teaching teams or with the appearance of observers;
3. I had open access to the classroom;
4. the teachers viewed team-teaching as a corporate learning experience and welcomed another pair of eyes to share the overview of the children's progress and to act as a reciprocal 'critical friend' concerning the assessment and analysis of teaching skills.

I wished to involve the children in exploring their own attitudes and perceptions toward underachievement but without suggesting responses by too narrow a question construct. The question posed was deliberately open-ended in order to encourage individual interpretation. This approach was also designed to elicit those responses that children found important in an attempt to sample a particular perspective on underachievement.

As a researcher I could pose the questions, respond to visual and verbal signals, share their environment and manage the context, but their experience was a matter only they could research:

> This approach calls for a very different form of teacher–pupil relationship and classroom climate . . . the role of the teacher is to set up a learning experience which encourages the pupils to reflect upon their own feelings, ideas and values. (Kyriacou, 1986)

Analysis of Responses

Student responses were divided into eighteen categories for the coding and the discussion follows the categorization. The main categories were The Home, The School, The Child and The Teacher; each main category had a number of subcategories (see table 11.2).

The home

28.4 per cent of the children's responses identified the home as being an important cause of underachievement. The younger children seemed influenced by insecure family relationships and personal trauma whilst older children seemed more aware of the debilitating effect of poverty and the depressing influence of too much family responsibility.

Rank order of children's perceptions as to the influences on achievement

1. Teacher's relationships with the child   15% of total
2. Child's relationships with others   12.3%

Other factors in rank order under the headings of:
A — the home 28.4%, B — the child 35.3%, C — the school 10.7%.
3. B Child's self-concept
4. A Relationships within the family
5. A Material facilities
6. C School discipline
7. A Parental attitudes to school
8. C Child behaviour
9. C Classroom climate
10. A Emotional disturbance within home
11. B Specific learning difficulties
12. A Family demands upon the child
13. B Personal medical problems
14. C Teacher's classroom skills
15. C Teacher's expectation
16. C School organisation
17. C School curriculum
18. C School environment

Parental attitudes to school

The children seemed to be aware of the negative influence that parents exerted by not displaying encouragement and interest in their achievements or in the development of good learning habits. The children were also concerned that their parents maintained a positive relationship with the school and showed support for success. Many children seemed to be aware of the importance of positive and encouraging attitudes and by the parents' relationship to the child's needs for success and approbation. But as well as material rewards the children testified to the desire to please their parents and gain approval by conforming with their parents' views of academic success. Quite often the fears and disappointments of the parents came through the comments made by the children and displayed the attitude that education represented for them. The attitudes of the parents just as often reinforced failure, both in their lack of support for the child and in their lack of interest in the school. The influence of parental attitudes on the development of positive learning habits remained constant throughout the four years under study.

Family relationships

This subsection received the largest number of responses in the category concerned with the influence of the home on underachievement. These responses represented 25 per cent of the total responses for this category and seemed to be more influential on the behaviour and achievement of the younger children than on the older children. The children saw aggressive and violent relationships as being contributory factors to the development of anxiety and the concomitant effect this had upon achievement. The younger children seemed concerned with a large number of responses relating to sibling rivalries. The older children seem to be far more aware of the more serious and threatening influences. Trying to cope with the breakdown of marriage, divorce, and death seemed to have a long-term influence on achievement. The influence could be traumatic enough to cause the development of school phobia and have a serious effect on the child's developing sense of identity. Some responses referred to the way in which the unconscious levels of guilt which were present in the breakdown of marriage were sometimes passed on to the children. Some were explicit and some were implied. The suggestion made by the parent, perhaps tendered subconsciously, was that the child was in some way a contributory factor to the problems within the home and that the child's performance and behaviour at home and school was somehow linked to the happy resolution of severe family difficulties. This implied threat and the child's guilt resulting from the inability to resolve the situation, could lead to high levels of anxiety and increasing performance and behaviour problems (Elton, 1989). These more serious problems seemed to remain unresolved and exerted constant influences on the child's attitude towards learning. They also influenced the way in which the child behaved in relationship to other children and the way in which s/he related to the school. In particular, they overreacted, sometimes violently, to rejection or censure.

Material facilities

The younger children seemed to be less influenced by material needs but were well aware of the influence of poverty, overcrowding and the lack of books in the home. By the fourth year, strong subcultural influences were modifying the attitudes of the children and there were many more responses that demonstrated the importance attached by the children to material things. The lack of material facilities did not seem to be serious in itself but increased the inability of the child to compete with others to be accepted as a member of a preferred peer group and to experience events and activities which others took for granted. This had implications for the way in which children evaluated their worth and for the development of values and attitudes.

Emotional disturbance

This was where events in the children's lives had caused deep-seated anxiety which were both long term and unresolved. The younger children seemed to be able to articulate these problems and were more willing to seek a caring teacher to share their concerns. The responses from the older children were mainly oblique references to threatening circumstances in their lives. They were less likely than the younger children to seek help, and were characterized by a simmering discontent.

Demands upon the children

Many of the children were used as substitute parents as they grew older. The problems of unemployment locally had resulted often in both parents having to take part-time employment, often during unsociable hours. Many responses referred to looking after younger children. It was noticeable in this category that the responses from the 12+ year (American grade 7) were more than six times greater than from the 9+ year (American grade 4). This had implications for the development of good learning habits, for the acquisition of poor role models and for the lack of adequate supervision.

The child

The largest number of responses related to this category of variables. They accounted for 35.3 per cent of the total responses. The 9+ year seemed more concerned about perceived learning difficulties. By 12+ many children had been diagnosed and remedial strategies developed. The younger children also seemed concerned with minor ailments and short-term medical problems and poor behaviour as major causes of underachievement. The older children were concerned about the way others perceived them, their relationships with others, and their relative worth within the classroom.

Specific learning problems

The children seemed quite aware of their learning difficulties, some of which may not have been diagnosed. Many children were only identified when they were referred for behavioural reasons. The number of responses dropped significantly from the 9+ to the 12+ year. Most of the younger children referred to relatively minor problems in general. Other younger children revealed how their feelings of inadequacy and failure led inevitably to frustration, a sense of failure and to antisocial behaviour. The responses from the 12+ year were fewer in number but represented a sad indictment upon the school, referring to few opportunities to succeed and lack of staff expectation. Already the sense of failure was well established and the children were caught in the trap of minimizing the importance of their failure in order to maintain a healthy self-image. There was also an undercurrent of sullen accusation which led inevitably to poor behaviour. Some responses questioned the sense of attending school if nothing was learned.

Medical problems

Most of the responses in this category originated in the 9+ year. The medical problems mentioned were relatively minor in nature and may be related to the nervous anxiety of moving from the lower school to the middle school. But there is a hard core of medical problems which, though they are not severe, interfered with the child's ability to develop positive learning habits. The more severe medical problems — diabetes, epilepsy, leukaemia and others — generally were well documented and received continuous and qualitative attention but they could still influence the child's ambitions, sense of achievement and view of personal worth.

Social

One prime motive for achievement appeared to be the attempt to earn status, esteem, approval and acceptance within the learning group. To some this meant achieving concrete evidence of success through healthy competition. Friendship groups were seen as positive elements in the acquisition of good learning habits. The responses in this section were far more numerous in the 12+ year and demonstrated a well-established, anti-establishment pupil subculture. Some pupils were caught in a vicious circle where their anti-school attitude and low attainment progressively reinforced each other. There was also heavy pressure brought to bear on children to conform to the anti-establishment views and there seemed to be a definite feeling that some children had little scope for achieving anything worthwhile on the school's terms. The children were driven to achieve a sense of belonging, of esteem and worth from membership of the counter-culture rather from belonging and identifying with the school ideals. Failing to win social approval through the curriculum and the school's formal activities they developed a well-defined set of behaviours and methods of punishing non-members, ranging from devaluing others' achievements to threatening physical punishment.

Behaviour

Whereas I had taken social relationships to be a manifestation of pupil to pupil behaviour, this section deals with classroom behaviour directed at the teacher. The responses ranged from simple non-compliance to overtly disruptive behaviour. The bulk of the responses from the 9+ year were relatively minor in nature and consisted of noisy or non-work related talkings, not getting on with the learning activity or mild misdemeanours such as being out of one's seat, fidgeting or eating. Though the number of reponses were fewer in number by the 11+ and 12+ years, the responses referred to much more serious behaviours. These included direct disobedience, physical aggression, refusal to accept discipline or punishment and persistent absence. Lying and stealing was a way of gaining esteem. Analysis of the responses from the older children indicated that the comments originated from ten separate groups and therefore the anti-school comments were related to about ten children who seem to have a disproportionate influence on the other children. The children also referred to boredom caused when the lesson content was of little interest because the learning activities were too passive or the manner of presentation failed to sustain their interest and attention. Poor behaviour was therefore related to achievement but was caused by a multiplicity of interrelated factors.

Self-concept

The recognition of the individual's worth seemed important to all years as was the need for self-realization. The children who were failing seemed to have the need to identify their failure as a physical attribute — and included the notions that being female, fat, black, etc., influence their ability to achieve academic success. The other aspect of self-concept seemed to be the worth that individuals placed upon themselves. If this involved a long period of an experience of academic failure then this seemed to undermine the development of a positive attitude and motivation towards learning. It seemed to contribute to a progressive alienation from school and what the school had to offer. The teacher reinforced the view of the child's perceived failure, often unwittingly, by the way he conveyed messages regarding attitudes, expectations about the child's ability and their behaviours. It was noticeable that though the children expressed public disdain, their written responses indicated the private importance they gave to the teachers' attitudes. The responses revealed that improvement of behaviour and attainment could be achieved by setting up experiences which enabled pupils to be more active in the planning, task and feedback of learning experiences. Also there seemed to be a tendency for children to look for a concerned teacher with whom to achieve individual and positive relationships.

School

The responses of the children intimated that in general they perceived the school as a supportive institution. They were more concerned with the curriculum organization which prevented them from avoiding staff that they disliked, distrusted or had no confidence in. There were only 10.7 per cent of the total responses referring to this variable and was the lowest response of all the categories.

Teacher

This category contained the variable with the highest number of total responses. The teachers' relationships with children received 69 responses representing 15 per cent of the total. This is perhaps a measure of the importance attached by the children to the influence of the teacher in encouraging or discouraging the achievement of children, the development of their education abilities, their socialization and their personality. Except for the younger children's more positive response to environmental conditions, the responses were evenly distributed across the variables and throughout the four years.

Skills

The children perceived this variable to have two distinct components; the teacher's general skills of delivery and the teacher's specific skills. The general skills received a great deal of comment and mentioned the audibility of instructions, clarity of blackboard work and organization of equipment. The children identified with specific skills those of setting appropriate levels of achievement, interesting content to lessons, teacher punctuality and appropriate feedback as being important.

Classroom climate

The younger children seemed more concerned with the physical layout and environment of the classroom and its contribution to a purposeful and exciting atmosphere. Good displays were the manifestation of the teacher's encouragement and demonstrated success and esteem for the children's work. The encouragement to work in groups and the informal arrangements of tables emphasized the active pupil role. The older children seemed to recognize this too but also were aware of how the teacher's demeanour, character, skills and abilities coloured the active life of the classroom. The responses saw the effective classroom to be where the authority of the teacher to organize and manage learning activities was accepted by the children. There was a mutual respect and friendly rapport. The atmosphere of a happy classroom was characterized by purposefulness and confidence in the learning process.

Relationships with children

This category received the largest number of responses of all the variables. The children identified the teacher's encouraging manner, fairness of relationships, good humour, patience and interest in individuals to be the most important components of the teacher's skill in encouraging children to achieve their potential. It was also felt to be an important quality in a teacher when he showed interest and exhibited encouragement in situations outside the classroom. Lack of sensitivity and sarcasm or derogatory remarks were disliked because the children could not defend themselves or reply in kind without being rude or disrespectful.

Expectation

This category received a very small number of responses. In many ways this was to be expected as it was a hidden aspect of the teacher's arsenal of skills. Clearly the expectation of the children must be related to their individual needs and must be encouraging as well as academically sound. Unfortunately, the underachieving child we have already seen (Kerry, 1982) is skilled at deception and difficult to diagnose. Many teachers did not concern themselves with children if they were tractable even if their abilities were not being extended. This led to boredom, irritation and eventually poor behaviour in an attempt by the child to attract attention to his difficulties. There seemed to be a self-fulfilling prophecy at work — that if we expected the worst of children, that is what we got!

When examining the responses to the research it was clear that the child perceived personal relationships to be very important. S/he was motivated by the way in which the teacher managed the 'hidden' classroom curriculum of values and attitudes and responded positively when the teacher displayed a genuine interest and curiosity about what pupils said and thought. It was this quality of professional concern for individuals and the development of reciprocal liking and respect which seemed most valued by pupils.

The relationships that the child developed with her or his peers and the status s/he enjoyed within the classroom was also perceived as an important influence on underachievement. Low-attaining pupils developed anti-school views and an anti-school 'counter-culture' developed as a response to a feeling that their achievements were not valued. An effective classroom experience was perceived as one that provided continual support and encouragement so that failure, when it occurred, was seen in itself to be a learning experience and did not undermine pupils' self-esteem regarding their learning:

> When dignity is damaged, one's deepest experience is of being inferior, unable, and powerless. My argument is that our secondary schools inflict such damage, in varying degrees, on many of their pupils. (Hargreaves, 1967)

Connected with the notion of self-concept was an awareness of teachers' expectations of pupils' attitudes towards attainment. Too high an academic level of expected achievement was seen as leading to a continuous and debilitating failure whilst too low was seen as boring and encouraged the children to develop an inflated and erroneous atitude about their own abilities. The children made much of genuine and helpful feedback to their work. They regarded rapid marking, positive comments and suggestions and a regard for the child's effort as good teaching qualities. The children also appreciated the opportunity to be given more control over their learning environment which enabled them to be more active in planning and conducting academic tasks and activities. The four top ranked variables in the research represented 45 per cent of the total responses to the total of eighteen variables:

- The teacher's relationships with the child
- The child's relationships (with his peers)
- The child's self-concept
- The child's relationships within the family

and seemed to suggest that the child regarded her or his performance as the response to a series of overlapping circles of influence and were related to how she was valued in her or his environment.

Comment

It is clear from the evidence that children consider the school to be a strongly positive influence in their lives, whilst the home is seen principally as representing negative influences. Of most importance to the children seems to be the way in which others value them as individuals and the relationship developed between teacher and child.

Research suggests that the background that the children come from, linguistically and socially deprived, has a negative influence upon achievement, but that a positive and encouraging learning environment can influence change (Steed and Lawrence, 1988). In this way, improvements in behaviour and achievement can be made as the school seeks to respond to the needs of the child.

The skills of the teacher in the development of relationships with the child are of particular importance. The child clearly sees these relationships as potential forces for influencing positive or negative achievements. Children see themselves as underachieving if their efforts are not valued and the learning environment is not encouraging and stress free. Further encouragement is perceived as coming from parental support.

The children see the school, its ethos and organization, as having a marked effect upon individual achievement by dividing the slow child from the bright and reinforcing the disadvantages of the weaker child. In order to maintain a positive self-image, these 'failing' children can become disruptive and anti-social, display serious learning difficulties, become backward at school and underachieve. Once this cycle of poor achievement and disruptive behaviour is allowed to become established it becomes increasingly difficult to modify. The school organization and the curriculum it offers, the discipline it exerts and the standards it sets can promote and reinforce anti-school attitudes and is seen by the children as a cause of underachievement. Where children perceive themselves to be rejected by the traditional school values they may seek to preserve their self-image and ameliorate high levels of anxiety by subscribing to a pupil subculture which glorifies failure and revels in poor behaviour. Once established, even clever and willing children say that they are unable to respond positively to the school ethos and the teacher can become increasingly isolated.

For some children underachievement was identified as being caused by poor motivation, poorly targeted curriculum objectives or poor teacher expectation. Disaffected children seemed to view the curriculum as uninteresting or irrelevant to their needs and constant failure became entrenched and perpetuated. However, in spite of the negative responses articulated, children also identified the school and the teacher as being the two areas that were most likely to give positive help. The children responded well to praise, reward and encouragement, well-targeted lessons and unthreatening classroom atmosphere. They see the 'good' teacher as being just, treating all children alike, being patient and good humoured. The badly behaved children seemed to realize their shortcomings but wanted the teacher to listen to their grievances, be given chances to reform and to develop positive relationships. In spite of overt evidence to the contrary, the children wanted the teacher to deal with the situation in which they felt trapped and powerless. They testified to being more willing to respond to interested teachers than to the impersonal discipline of the school.

The children articulated the fundamental value of positive personal relationships between the teacher and child in the development of a successful learning process. The teachers involved in the research began to realize that positive action could result from the recognition that far from being powerless in the face of underachievement, it was possible to develop strategies to break the cycle of disaffection, deprivation and underachievement. The variables influencing underachievement could not solely be attributed to factors outside the control of the school.

Schools Can Make a Difference

The concept of inservice education also being collaborative action research is a useful paradigm for other schools to use. The teachers taking part in the research were impressed by the way in which targets related directly to the needs of the school, and were capable of modification or response to emerging issues. Economically the cost was minimal as the research sought to use the expertise of the staff. External courses, visiting experts and 'patch' initiatives could be fed into the research as the need arose.

The classroom support of teachers through team-teaching initiatives was identified as particularly helpful in generating and sharing resources and supporting discipline. It also enabled good practice and skills in teaching to be shared through the 'mentoring' practice of pairing young teachers with more experienced colleagues.

The research also used existing school management and administrative systems in order to minimize unnecessary change and to identify and strengthen good practice. In this way change was perceived as responding to need and was rooted in the corporate requirements of the children and teachers. The ideas of individual staff were valued, a collaborative 'ownership' of the research encouraged, and a cycle of identification intervention, development of resources, assessment and identification, developed.

Underachievement seems to be related to the school atmosphere, as well as the other factors discussed earlier. An effective and positive learning environment seems to promote good levels of educational achievement and behaviour. This does not mean that the school can eliminate the influence of social differences. The research suggests that disadvantaged children still perform, on average, less well than children from an advantaged home. However, a disadvantaged child in a positive learning environment is likely to do better than an advantaged child in an ineffective learning situation. It can be said in broad terms that an underachieving teacher is the principal cause of underachieving children and that an underachieving teacher is the result of an underachieving school.

What is clear is that the school routine — timetables, standards of work, behaviour and attendance — work together to produce an effect on children's behaviour and attainment. The teacher's attitude and response to the standards of behaviour, for instance, seems to influence the way in which the children behave. An ineffective teacher's classroom, for instance, was characterized by:

- widespread litter in the classroom;
- graffiti on the desks;
- ignoring bad behaviour in the playground and corridor;
- work not displayed and marked infrequently;
- pupils late and getting away with it;
- a lack of politeness;
- erratic and inappropriate punishments;
- lessons not prepared;
- frequent confrontations;
- very little humour.

as articulated by teachers in their year group meetings. This seems to contribute towards a lack of cohesiveness and a sense of community within the classroom. Teachers are under great stress and feel isolated and the children feel undervalued; they are expected to achieve poorly and behave badly, and so they do. Effective action starts with the recognition that underachievement cannot simply be attributed to factors outside the school but that it may be the result of the school's failure to identify difficulties and provide a positive learning environment.

The way in which schools are managed and classrooms organized can be changed. This can be an uncomfortable process if teachers feel threatened. The first priority is the recognition of the need for change and the second a positive commitment to see it through. If imposed from the top by a linear system of command, change was seen to be autocratic and often ineffective in carrying the staff along. What was crucial was that teachers were clearly focused on the problem and developed a corporate ownership in change. It was seen to be related to the needs of the teachers and children and that the teachers' ideas could modify the process and that the process could respond to emerging issues. In this way the development of teamwork, the encouragement of a collective responsibility for the curriculum and improved communication and consultation, focused on valuing and supporting teachers in the process of change.

The effective management system in this situation is a combination of line management and the consultative role coupled with the 'first among equals' philosophy. Teachers repeatedly indicated during the research that:

1. they needed to feel that their school had a sense of direction;
2. they needed to feel that this direction could be influenced by their views;
3. they needed to feel valued as individuals in this process.

References

BERNSTEIN, B. (1971) *Class, Codes and Control*, London, Routledge.

BULLOCK, A. (1975) *A Language for Life (Report for DES)*, London, HMSO.

COLEMAN, J.A. (1993) *Equality and Achievement in Education*, West View, Allbright.

DES (1989) *Discipline in Schools (Elton Report)*, London, HMSO.

DES (1985a) *Curriculum 5–16 (HMI Report)*, London, HMSO.

DES (1985b) *Education Observed: Good Teachers*, London, HMSO.

ELTON REPORT (1989) *Discipline in Schools*, London, HMSO.

HARGREAVES, D. (1967) *Social Relations in Secondary Schools*, New York, Minnesota Press.

KERRY, T. (1982) *Teaching Bright Children in Mixed Ability Classes*, London, Focus Books.

KYRIACOU, C. (1986) *Effective Teaching in Schools*, Oxford, Basil Blackwell.

McNAMARA, D.R. (1980) 'The outsiders arrogance: A failure of educational research', *Participant Observers*, **6**.

MORTIMER, P. (1988) *School Matters*, London, Open Books.

SATTERLEY, D. (1985) *Assessment in Schools*, Oxford, Basil Blackwell.

SLAVIN, R.E. (1991) *Student Team Learning*, London, National English Association (NEA).

SPIELBURGER, C.D. and GAUDRY, E. (1974) *Anxiety and Educational Achievement*, Milton Keynes, Open University Press.

STEED, D. and LAWRENCE, J. (1988) *Disruptive Behaviour in the Primary School*, London, London University.

STOTT, D.H. (1976) *Social Adjustment of Children: Manual to the Bristol Social Adjustment Guide*, London, London University.

SWANN, M. (1985) *A Response: (A Report Produced for the Committee of Racial Equality)*, London, HMSO.

THORNDIKE, R.L. and HAGEN, E.P. (1969) *Measurement and Evaluation in Psychology and Education*, Chichester, Wiley.

TOWNSEND, H.E.R. (1971) *Immigrant Pupils in England*, Windsor, NFER.

WEST, D.J. (1982) *Delinquency: Its Roots*, London, Heinemann.

WHITAKER, P. (1984) *World Studies and the Learning Process*, World Studies, Journal 5, Paris, OECD.

12 Emotional Experiences in Learning and Teaching in a Distance Education Course

Rosemary Crowe

I have always been interested, fascinated in fact, by the way we emotionally experience teaching and learning. We have all been through the extensive socializing process of primary and secondary education and, for all educational professionals, we have continued through other institutions to further areas of study and experience. Then for the most part we have been on the other side of the fence as teachers and educators. In all of these experiences, each and every day there is an element of emotion and emotive response to particular episodes, events, other people, subject material, sometimes even the physical surroundings in which we have found ourselves.

In this paper I am discussing the experiences of a cohort of students in Hong Kong who are studying in distance education mode, through Deakin University. The course is built around a group of units of study which have strong critical reflection elements and much of the assignment work that students complete is based on an action research model. Students are invited throughout all the units to offer evaluative comments and reflections on their own work and the units of study themselves. The students use this forum as an opportunity to make comments and offer insights about all elements of the units they are working on. These vary from the quality of the materials, support from tutors and organization of the course to the effect of what they are learning on their own practice, on the children in their classes, on their personal lives and on themselves. Many of these comments have been strongly focused on the affect and feeling the course work has resulted in and it is that strong sense of emotion that has made me as course tutor and coordinator really respond as well, on a deeper level.

Students are, most definitely, moving through fundamental changes as they work through the course. These are changes which are different to those imposed perhaps by economic, political or organizational change although there is certainly evidence of those elements in student writing. Instead, they seem more fundamental and more intense to some degree as students change their own practice, their own way of looking at their schools and classrooms and themselves as teachers and students.

The amount of responses students gave when they wrote comments in an evaluative sense began to show a pattern where the strong personal responses were more common than those which were more formal. I realized that I, as the teacher in this situation, was experiencing strong emotional responses to my work and to the relationship I had with these students. As I started to explore some of the work written about emotional experience in learning and teaching I could recognize these responses of both the students and myself in the literature.

Emotion is not a word that is used often in teacher education programs nor in our every day dealings as educators or students. It is integrally there, however, as an element in all our personal dealings with others and within our experiences as reflective adults. A strong reason as to why we do not use the word emotion very often could well be the meaning some give to the use of the word. It evokes responses which are not all together reasonable; the most common issue surrounding emotion, which is also one of the oldest, is the opposition between emotion and reason. Originating in philosophy and Christian theology, emotion is viewed as 'passionate', as out of our control and in direct contrast with reason. Early Christian religious-based philosophy regarded reason as that function which was within our control. The opposition of reason was passion or emotion is paralleled by the opposition of male and female where, 'men are seen as rational and women emotional, lacking rationality' much of this belief can be traced through western thought (Crawford, Kippax, Onyx, Gault and Benton, 1992).

Much psychology with its emphasis on objective scientific approaches in turn, while not being able to come to terms with the idea of rationality and emotionality being in opposition and not able to co-exist, ousted the mind in favour of the body as the focus for psychological study. It was then to the body that psychologists turned when they came to explore emotion and the physiological responses experienced were explored to try and explain emotion. The authors of *Emotion and Gender* (Crawford, Kippax, Onyx, Gault and Benton, 1992) which I have used here offer an interesting and critical overview of the accepted theories of emotion taking the stance that emotion is socially constructed.

So why talk about emotion in conjunction with action research? As I have already stated, much of our teaching life involves responses to situations which are emotional responses. Feelings and affect which can be intense or minimal are involved in the experience of emotion. The experiences of both students and tutors which I discuss here are integrally related to change. If we accept the following about the nature and characteristics of change, then we are also accepting a high level of emotional response. Change is:

- about dealing with perceptions and feelings;
- about people more than things;
- likely to cause conflict;
- dynamic;
- about process more than product. (Smith and Lovat, 1990)

There is much to understand in this area of study. I hope that through some reflection on what are some quite common responses to a unit of study in a distance education mode, fruitful discussion will ensue as to how we can further enhance practice which will enable richer experiences for both students and teachers.

The Students

This group of students are completing a Bachelor of Education program in Hong Kong which is offered in a 'twinning arrangement' through the Chinese University of Hong Kong and Deakin University. Students are qualified teachers who wish to upgrade their qualification to BEds. Students have completed either a two or three year teacher education qualification and all have completed English at O-level standard. Most teach in the medium of English. Students entering the course are required to complete ten units of study on average. Some with slightly less previous study need to complete twelve units and a few need only complete eight units of study. Nearly all students are full-time teachers who work in regular schools, although some are placed in special schools, and a small number are involved as trainers in industry or teacher college education.

These students have been educated in a system which could be considered a broad reflection of the old grammar school system of Britain. Ninety per cent of students in Hong Kong attend Anglo-Chinese schools and officially teaching is in English (Morris, 1990), although one has to consider how difficult this is when very little contact is made with people who have English as their first language. Places are highly competitive in the streamed education system of Hong Kong. Students compete to enter the more prestigious Band 1 and 2 secondary schools which are highly academic and those children of wealthy families leave the colony altogether to complete their secondary education in the United States, Canada, Britain and Australia. It is very much part of Chinese tradition to highly respect the scholarly and rigid examination system. Hong Kong's education system is based in a traditional colonial role and one that now is in the unique situation of transition to a socialistic political ideology.

The Course

These units are the same units that are offered in Australia and the course materials remain the same. Some texts and readings that are more appropriate to students in Hong Kong have been incorporated into the units of study. The units of the course are designed as a package. Unlike the off-campus students completing the degree in Australia who have the choice of the wide range of BEd units of study. A group of units has been selected and presented in an order to allow the formation of an integrated course. These units are two

Sociology units, Classroom Processes, Changing Curriculum, Evaluating Children's Progress, Educational Enquiry and a Supervised Individual Project. The
three latter units of study all require students to undergo a specific process of
change which is based on an action research model as described by Kemmis
and McTaggart (1982) where

> Action research is a form of self reflective enquiry undertaken by
> participants in social situations in order to improve the rationality
> justice of their own social or educational practices, as well as under-
> standing of these practices and the situations in which the practices
> are carried out. (p. 5)

The Unit of Study

The unit of study that these particular personal evaluation statements were
gathered from is Changing Curriculum. This is a unit of study which, as the
name suggests, brings together the complex integration and interrelation of the
many aspects of curriculum at many different levels. It also considers the idea
of change. Curriculum is, by definition, about change: about changing circumstances and changing people, and about changes in terms of our understandings
of ourselves and our world. Not only is curriculum about changing things in
this way, however, but also it is itself changing always and incessantly because
it exists in history and therefore cannot help but be caught up in the historical
process itself (Study Guide, 1992). As this exemplifies, this is a complex unit
which regards the notions of curriculum and the notions of change against an
holistic framework of 'macro', 'meso', 'micro' relationships. The writers of the
course use the metaphor of curriculum as story and encourage students to
consider their work through the course as that of a journey which can be
undertaken in a myriad of different ways. Students use the *action* research
model to change their practice in a particular curriculum area in some way.

The unit is organised into key sections, and in each section a number of
tasks, not exceeding nine in each key section, constitute the assignment requirements for each section and are graded. Tasks draw on specific readings,
or materials or ask for students to compile information of their own.

It is the openness and the use of the strong metaphors of journey and
story which invite and encourage students to critically reflect their own experience of the work they do as they move through the unit of study. Alistair
Morgan makes the case that a basic tenet for improving students' learning is
to take account of their experiences. The essence of reflection he considers is
to 'explore experiences and move on to new understandings' (Morgan, 1993,
p. 124). It is considered, then, an important element of the students' learning
experience, that they take time to genuinely reflect and consider their own
work and understanding of the course.

Evaluation Opportunities

Enabling students the opportunity to respond to a course structure through a more personal means gives the opportunity for a dialogue which is less concerned with the formal assessed work of the course and more concerned with: what the experience is for the students; how they are coping with it; what suggestions they may have to offer in the evolution of the unit; whether students are receiving from the course that which the writers of the unit envisaged. These are very wide ranging evaluative questions that can find answers in the personal, reflective writing of those taking part in the unit.

Students were told the following about Personal Evaluation Statements in the study guide for the unit:

> These usually occur at the mid-point and end of each section. They are not part of your assessment, but are included to enable you to respond openly to the unit as you move through it. We place high value on your responses which are central to our efforts to refine and improve the unit from year to year. (Unit Study Guide, 1993, p. 2)

The following statements were part of the task structure:

> Please take a few moments to write down your feelings about the tasks and the course so far.

> . . . now you have finished the first leg of the journey, we would welcome any comments particularly any suggestions on how we could improve this section of Changing Curriculum.

We do often ask for students to write some sort of personal response to a particular study experience, but generally speaking little use is made of these responses other than to perhaps evaluate a course on a wider level of response or to justify the course to those who have designed it.

What does appear most strongly is the emotional aspects to learning that can be found in many, many learning situations coming through the writing of this group of students. These bring in aspects of the relationship between student and teacher, the aspirations of the teacher as well as the student and the difficulties of taking the special responsibility for our own learning that is inherent in distance education units of study.

Beginnings

A useful way to explore this is to consider the process of a unit of study through the areas of its beginning, middle and end. Salzberger-Wittenberg explores these notions through the hopefulness and fearfulness of the expectations we all share when beginning something in our lives that is important. I have tended to draw heavily on the ideas of Salzberger-Wittenberg (1983) for the purposes of looking at these responses. Her book which aims to heighten the awareness of the emotional factors which enter into the process of learning and teaching offers an outline which can be used as a framework for taking these sort of responses and examining them in a way that allows for greater understanding of what the student is experiencing. This greater understanding, too, allows for reflection on the teacher's part to make more sense of the learning experience of the student and the experience of being a teacher.

In the early stages of a unit of study, students do often feel overwhelmed by the sheer enormity of the task they are facing and the changes that they can see themselves working through in their teaching settings. This is often felt more intensely in a distance education unit where all the materials arrive at once, with all the instructions and assignments for the course and the student is alone in their own home trying to make sense of the package that has arrived. The very first task in this unit is to spread the materials out on the floor and consider the contents in an overall manner.

> It is the nature of beginning that the path ahead is unknown, leaving us poised as we enter upon it between wondrous excitement and anxious dread. (Salzberger-Wittenberg, 1983, p. 3)

This hopefulness and fearfulness finds expression in many of the comments made by students in their first notes of personal evaluation: feelings of inadequacies to cope with the course itself; the sense of being overwhelmed that can effect many of us at the beginning of an important task; because of the sheer quantity of the material to be considered as part of studying a course. Salzberger-Wittenberg (1983) discusses these feelings within the framework of their infancy roots, putting forward the idea that any new idea reawakens the feelings of loss, helplessness and confusion that we all dread. One student wrote that her feelings, at this early stage, could be best expressed through the following which she shared with us:

> There is something I don't know that I am supposed to know,
> I don't know what it is I don't know and yet am supposed to know,
> And I feel I look stupid, if I seem both not to know it, and not know what it is I don't know,
> Therefore I pretend to know it. This is nerve-wracking since I don't know what I must pretend to know.
> Therefore I pretend to know everything.
> ('Knots', R.D. Laing, 1979)

This is a wonderful poem in its expression of all those worries and emotions that do beset, probably, many of us at the beginning of any new and challenging task. I have chosen it to represent a good many of the comments and perceptions of other students in this group as shown in their personal evaluations at this early point in the course. It was not universal, of course, this talking of the anxiety of as Salzberger-Wittenberg says, 'being in a state of beginning once more'. What effected me so strongly reading this was it described so well how I was feeling. I was new to the course as tutor and I was feeling the stress that students were feeling myself.

The metaphor of journey and story through the course is a strong one and also offers the students a metaphorical base to use themselves. This student put it eloquently in the first personal evaluation completed where he chose a very strong focus on the metaphor of journey:

> I believe that I am progressing towards the 'storehouse' and yet the route would be difficult with hurdles along the way. Like the pilgrim I have gone forward only to fall back. Like him, I have fallen back only to press on once more. The tasks to me are my burden to shoulder. On some occasions I have laid it down for a brief respite only to be obliged to pick it up again and find it even heavier. So I realise that I should keep my eyes fixed on the destination — the far off Delectable Mountains of good will and so I go on my journey — the pilgrimage.

> When I first get this set of materials, I feel frightened because it seems it must take a lot of time to prepare the assignment. Besides, when I compare this set of materials with the first one I studied, it seems that it is more complicated and difficult to study than the first one. Also, the word curriculum, in some ways I am quite familiar with it, but in another way it is strange to me. Up to now I have read a lot of materials, I have done some tasks required, I have learned something about the curriculum, i.e., definition of it. The tasks or the contents give me implication of education system or curriculum. I begin to gain interest in Hong Kong's education curriculum or system. I hope I can do my best in my remaining tasks in order to gain more insights in it.

Student Expectations

The same student says in a later statement in this period of time that:

> . . . more abstract ideas have come to me. However I wonder if what I grasp is right or wrong . . . I just hope my work won't let you down.

The student has struggled with the ideas herself but is quite used to completing college work and before that secondary school work where the material was read, absorbed and to a greater extent reiterated in the best way possible to gain a good result. The agenda was clear and the student knew what to pursue. In this case the agenda is not so clear. It is rather new and challenging.

There is a strong sense of worry from some students that they are not on the right path and that in this regard feel uneasy because they are not sure what it is that the tutors are looking for. With these students there is also the added difficulty that English is not their first language. While on the most part English is excellent and it is the language in which many were taught at school and in which many of them themselves teach, the reading offered in this unit and in the course overall are by their very nature challenging readings. Students have often taken the route of choosing to concentrate on a lesser number of readings more fully than to try and cope with the whole of the materials offered. Comments regarding this element of feelings about the course run in this vein:

> The demand for the students in the course is so heavy and as a new comer in the curriculum field, I don't think that I have a good ability to understand and to accommodate such a complicated course. The course is approaching in a way that I am not used to. It is a completely new and difficult course. I do hope that what I have done fulfils the aims of the tasks and the construction of my own story is in the right direction that the course wants me to go.

The next comment comes from a student who did particularly well in all the tasks of the course through an incredibly thorough approach to the work of the unit.

> There are various tasks to guide us to travel along the journey. I sometimes feel that the arrangement of the tasks is not so clear. I suggest the guidelines in the task and the aim of the task are written separately so that I can easily identify what the things I need to write to reply.

This is an area that is constantly mentioned by a number of students. They find the non-authoritative approach of what and how particular work will be undertaken very difficult to deal with. Students from many different learning settings want this in the relationship they have with the teacher/tutor. They want to know exactly what it is the teacher desires of them rather than feeling confident to pursue their own ideas through the course. One can assume this is integrally related to the system of marking and grading that we, and nearly all tertiary settings have. It is also part of the socialization that we have experienced as students in primary and secondary school itself.

Even though the unit of study is built around critical reflection and the changing of curriculum ideas within the student's classroom through a research process, it is very hard for some students to take the ownership which is offered. It is, however, an enormously difficult concept. It is realistic to expect that the teacher will have more knowledge than the student in a particular area of expertise and there is an insistence on the part of the student that the teacher will impart all necessary information and exactly how it should be treated by the student. Salzberger-Wittenberg (1983) states:

> I have often found students feel angry and cheated as if one was withholding something they feel entitled to and could possess if only the teacher would be more willing to share. (p. 26)

I, in turn, felt as the tutor on the course that perhaps I was not 'giving' or 'caring' enough at this stage. In fact I really did begin to feel a burden of guilt. When researchers talk about teachers' emotions they often talk of pride, commitment, uncertainty, creativity; teachers themselves can often talk about emotions like frustration, anxiety and guilt.

'Guilt is a central emotional preoccupation for teachers' (Hargreaves, 1994). Hargreaves describes those who work in the caring professions as being prone to depressive guilt or that guilt that we can experience when we realize we may be neglecting in some way or even harming those for whom we care, perhaps by not meeting needs or not giving enough attention. This could well be one of the integral emotions which we feel as teachers.

The open-endedness of teaching is another guilt trap of teaching. It is in reality a 'never-ending story'. There is always something more to do, something more to prepare, something more to think about. Teaching is defined broadly in terms of social, academic and emotional parameters.
I actually wrote this at this time in my own journal:

> am I feeling guilty because I have so much to prepare and so little time?
> am I feeling guilty because I do not know how much to give and I never feel like I have given enough?
> am I feeling guilty because I have paid more attention to administration than the course itself? am I feeling guilty because this student's work is really not up to scratch and I feel like I could have done better in my actual teaching of the subject?

Hargreaves also discusses the impossible ideals of perfection that many of us, as teachers, place upon ourselves. What I found here was that this was certainly an area that students were feeling acutely as well. Could it be, as students and teachers, they were experiencing a mixture of all these emotions in a different way than someone who is strictly in the role of teacher or of student?

Quite a few students have expressed anxiety at needing to 'please' the tutor. The expectation that the teacher will hand 'it' over, it being the body of knowledge and the way it should be done. This especially is a unit of study in which passivity on the part of the student is not rewarded in any way, where individual paths are meant to be taken by the student and where it is almost impossible for the tutor to 'magically' pass on the knowledge of the course and the way the tasks involved should be approached.

Many of these students hold firm the belief that their task as teachers is to transmit knowledge, that the children they teach are the proverbial 'empty vessels' awaiting filling with the appropriate educational material and values. This belief is reiterated again and again through this unit of study and others where the evaluation of students is considered. Comments such as 'the teacher's job is to transmit the knowledge to our students in school through the assigned and set syllabus' are very common through the work of these students. A unit of study which actually challenges this notion is often difficult and does meet with some resistance. Emotionally, it definitely seemed as though teacher and students were experiencing either side of the same coin. The question then arises, is this a regular happening?

Changes in Perceptions

Much of what has been discussed has come from assignments early in the unit. The students were feeling many of the emotions easily identified with beginnings and learning something new. An assignment later, the third in the series of four, comments began to change in nature.

In this section, also, readings were included which were specifically related to the Hong Kong experience. The students overall found the work more interesting and relevant to their own teaching settings as they were able to look at their own role in the 'meso' perspective. Which was the key focus in this segment of the course. One student's personal evaluation statement was made in this way:

> I have been teaching for five years, I have cooperated with the school and the system, I have taught to my students with the system and the school frame. Although I enjoy much in it, I haven't considered much on my role, responsibility, participation and autonomy in decision. It can help me to think better. Also, in this school, it challenges me to carry out a degree of self-analysis and reflection into my own teaching.

At this stage of the unit of study, students were very much involved in the work of the unit and tended to consider the reading and work they were doing in respect to their own teaching and work settings. Students seemed surprised overall at the autonomy they found themselves being able to take in their own classroom. They were not so much wanting to resist the system, but to take control and ownership of their work within their own setting. Leung (1990) suggests the most ready explanation for this is traditional Chinese culture and views today's political culture as an inheritance of social apathy and acquiescence. Can this be generalized into teachers' lives? Many teachers as students of this course report the heavy burdens of set syllabi and curricula which leave little ground for change. The existence of an external examination system which is highly regarded and respected by most leaves little room for negotiation of the curriculum by teachers.

The questions about 'meso' are very important to me. It is because it refers to matters about teachers, the area is about me. I can enlarge my knowledge from these tasks. I did learn a lot in doing all the tasks. I have another way to see things. Though I cannot reflect my words to the Education Department, I can make decisions about myself in the classroom. I have gained back some confidence.

Students made many mentions at this stage in their evaluations of gaining more confidence in their own classroom and feeling to some extent so empowered by what they had read that they were able to make decisions in the classroom for themselves. They saw this area of the unit as being very much to do with them as teachers personally. There was a certain fear expressed through some writing that it was necessary to be conspicuous or to stand out within the school structure in order to elicit any curriculum change, no matter how small:

Classroom teaching should be considered in a broader educational sense. Teachers should increasingly take initiatives to analyse, share and improve practice. They should be willing to overcome problems and to take initiatives to introduce and lead changes, though frequently this may require them to give up practices in which they feel secure and display high levels of competence.

Finishing

As students came to the end of the unit they all wrote of being more comfortable with the concepts of the unit and felt that the final two key areas had actually been easier than the first. There was certainly a sense of pride and accomplishment at having completed what had begun as a particularly daunting unit for some students. Evaluative statements at this stage described the attempts of students to use some of the curriculum models they had learnt about through the unit. Not all of these were successful attempts but there was a feeling of endeavour and initiative and the wish to test things out. Some students noted a real change and development in the understanding they had of their own school practices. It was suggested by more than one students that they had just never been able to look at it in that way before. A final comment from one student:

This is a very practical section for teachers. It helps me to think over MY teaching in a new way. The different models enriched my knowledge of classroom teaching and provided me with more information in planning a lesson. I think this is also a summary section of this unit on changing curriculum. Task 4.4 gives me the chance to recall what I have learnt and reorganize the previous knowledge in a systematic way. On the whole I have learnt much.

It appears most important that students are given the opportunity to address what they have learnt through their own practice and to bring what they have learnt to a careful and summarizing conclusion. Comments were overwhelmingly positive at this stage. One could assume that the relief of finishing a heavy task would be the natural answer to that, but it was with a sense of purpose and value that most students seemed to complete the unit:

> I learnt that I should be courageous to introduce innovations to curriculum and pedagogies most inducive to the kids.

Salzberger-Wittenberg (1983) speaks of a sense of loss that many students experience at the end of any particular area of their education, be it the end of a unit such as this. My overwhelming belief was that for these distance education students studying under quite different circumstances to many others of our students, the experience was enlightening, and empowering. By encouraging our students to engage in this dialogue of personal reflection we have been able to be a part of that and to be aware of it. It is far more than merely justifying our own abilities to design courses, it allows us an insight into the real experience of our students and to perhaps consider those elements which have, in this experience, proved to be the most stressful and difficult.

What does this strong sense of emotion tell us as researchers in education, as collaborators? I find it difficult to come to strong conclusions, especially in a paper such as this which has so briefly touched on many important issues, ranging from study at a distance, cross-cultural study, to the theme of emotion in response to learning and teaching. Perhaps I would like to finish by saying that the way we feel when we study or teach must be taken into account in an important way, not in a peripheral way. That our feelings and the affect of these experiences be counted into the overall experience by us all.

References

COURSE TEAM (1993) *Changing Curriculum, Unit Study Guide*, Geelong, Deakin University.

CRAWFORD, J., KIPPAX, S., ONYX, J., GAULT, U. and BENTON, P. (1992) *Emotion and Gender*, London, Sage.

HARGREAVES, A. (1994) *Changing Teachers, Changing Times: Teachers Work and Culture in the Postmodern Age*, London, Cassell.

KEMMIS, S. and Ms TAGGART, R. (1982) 'The Action Research Planner Geelong', Deakin University Press.

LEUNG, B.K.P. (1990) 'Power and politics: A critical analysis', in LEUNG, B.K.P. (ed.) *Social Issues in Hong Kong*, Hong Kong, Oxford University Press.

LAING, R.O. (1979) Sonnets, London, Joseph.

MORGAN, A. (1993) *Improving Your Students' Learning*, London, Kogan Page.

SALZBERGER-WITTENBERG, L., HENRY, G. and OSBORNE, E. (1983) *The Emotional Experience of Learning and Teaching*, London, Routledge.

Notes on Contributors

Clem Adelman is Professor of Education at the University of Reading, UK.

Sonia Burnard and **Heather Yaxley** are teachers in the Nazeing Park School, Essex, UK.

Sue Cox is a Senior Lecturer in Primary Teacher Education at the Nottingham Trent University, UK.

Rosemary Crowe works at the Faculty of Education, Deakin University, Geelong, Australia.

Anne Edwards is a Professor in Education at Leeds University, UK.

John Elliott is Professor of Education at the University of East Anglia, UK.

Sandra Hollingsworth is Associate Professor in Pre/Inservice Education at Michigan State University, USA.

Christine O'Hanlon is a Senior Lecturer at the School of Education, University of Birmingham, UK.

Alice Paige-Smith is a Senior Lecturer in Education, University of Hertfordshire, UK.

Peter Posch is a Professor of Education at the University of Klagenfurt, Austria.

Jane Richards teaches writing courses at the State University of New York College at Cortland, USA.

Melanie Walker is Director of the Academic Development Centre, University of the Western Cape, South Africa.

David Wilcockson is a Senior Lecturer in Education at the De Montford University, UK.

Index